Teaching with Tasks for Effective Mathematics Learning

MATHEMATICS TEACHER EDUCATION

VOLUME 9

SCOPE

The Mathematics Teacher Education book series presents relevant research and innovative international developments with respect to the preparation and professional development of mathematics teachers. A better understanding of teachers' cognitions as well as knowledge about effective models for preservice and inservice teacher education is fundamental for mathematics education at the primary, secondary and tertiary level in the various contexts and cultures across the world. Therefore, considerable research is needed to understand what facilitates and impedes mathematics teachers' professional learning. The series aims to provide a significant resource for teachers, teacher educators and graduate students by introducing and critically reflecting new ideas, concepts and findings of research in teacher education.

For further volumes:
http://www.springer.com/series/6327

Peter Sullivan · Doug Clarke · Barbara Clarke

Teaching with Tasks for Effective Mathematics Learning

Springer

Peter Sullivan
Faculty of Education
Monash University
Clayton, VIC, Australia

Doug Clarke
Mathematics Teaching and Learning
Research Centre
Australian Catholic University
Fitzroy, VIC, Australia

Barbara Clarke
Faculty of Education
Monash University
Frankston, VIC, Australia

ISBN 978-1-4614-4680-4 ISBN 978-1-4614-4681-1 (eBook)
DOI 10.1007/978-1-4614-4681-1
Springer New York Heidelberg Dordrecht London

Library of Congress Control Number: 2012945394

Printed on acid-free paper

Springer is part of Springer Science+Business Media (www.springer.com)

Acknowledgements

This project was funded by an Australian Research Council Linkage Grant, and we are grateful for the support of the ARC. We also acknowledge the contribution of the Department of Education and Early Childhood Development (DEECD, Victoria, Australia) and the Catholic Education Office Melbourne (CEOM, Australia) who assisted with funding, provided links with schools, and offered helpful guidance throughout the project. In particular, we thank Nadia Walker and Helen Gist (DEECD) and Gerard Lewis and Janeane Anderson (CEOM) for their roles in this.

We thank the three clusters of schools and their cluster leaders, Merging Minds Cluster (Jan McCluskey): Armadale Primary School, Hawthorn Secondary College, Lloyd Street Primary School, Malvern Central School, and Toorak Primary School; Berwick South Cluster (Peter Sanders): Berwick Fields Primary School, Brentwood Park Primary School, Hallam Valley Primary School, Hillsmeade Primary School, Kambrya College, and Timbarra Primary School; and Geelong Cluster (Peter Caddy): Our Lady Star of the Sea Ocean Grove, St Aloysius Queenscliff, St Anthony's Lara, St Ignatius Secondary College, St Margaret's East Geelong, St Therese Torquay, and St Thomas Drysdale.

There was a group of teachers within the clusters who volunteered to develop and teach extended units of work as the research team studied their teaching. We are particularly grateful to these teachers. They were Shaye Bradbury, Elizabeth Campbell, Anthony Cavagna, Robyn Frigo, John Hassett, Alison Jordan, and Chris Powers.

We also thank Helen O'Shea who was a Senior Research Fellow on the project for the last two years and who made a major contribution to the data on students which are reported in this book. We acknowledge too the work of Anne Roche in data collection, professional development leadership, and research assistance. We also thank Orit Zaslavasky, Ron Tzur, and Irit Peled who contributed to the project when visiting Monash University. Meredith Begg and Leicha Bragg assisted with the documentation of lessons and the development of the website, while Alison Jansz-Senn and Nike Prince undertook important administrative and research work in the earlier years of the project. Wendy May assisted us greatly in bringing this book to fruition with her management and editorial skills in the later stages, while Sheryl Sullivan assisted with refining the tasks included in the final chapter.

Sarah Ferguson and Richard O'Donovan completed PhDs which focused on aspects of the project involving low attaining students and the challenges teachers experience with using tasks effectively, respectively.

About the Authors

Barbara Clarke is an Associate Professor in Mathematics Education at Monash University (Peninsula Campus) where she teaches primary pre-service teachers. Barbara has considerable experience in conducting and supporting research and directing or supervising major research projects and contracts. The major focus of her writing and research has been concerned with mathematics teachers, their practice, and their professional development.

Doug Clarke is a Professor of Mathematics Education at the Australian Catholic University (Melbourne), where he directs the Mathematics Teaching and Learning Research Centre. In recent years, Doug has worked on four Australian Research Council grants, focusing on integrating mathematics and science, the role of tasks in mathematics learning, encouraging student persistence while working on challenging tasks, and providing appropriate support for teachers implementing national curricula, respectively. Doug's professional interests include young children's mathematical learning, using mathematics to explore current events and students' interests, the role of task-based assessment interviews with students, problem solving and investigations, manageable and meaningful assessment, and the professional growth of mathematics teachers.

Peter Sullivan is a Professor of Science, Mathematics, and Technology at Monash University. His main professional achievements are in the field of research. His recent research includes four Australian Research Council grants funded projects. He is an author of the popular teacher resource *Open-ended maths activities: Using good questions to enhance learning* that is published in the US as *Good questions for math teaching*. Until recently he was chief editor of the *Journal of Mathematics Teacher Education*, was immediate past president of the Australian Association of Mathematics Teachers, was the author of the Shape paper that outlines the principles for the development of the Australian Curriculum in Mathematics, and was the author of the 2011 *Australian Education Review* on research informed strategies for teaching mathematics.

Contents

Chapter 1
Researching Tasks in Mathematics Classrooms

This book is about the selection and use of tasks that can facilitate the learning of mathematics. It focuses on the choices of tasks that teachers make and the actions they take in using tasks. It describes tasks that can be used to facilitate the learning of mathematics, and can make the experience of learning mathematics engaging and rewarding and teaching of mathematics fulfilling. It draws on findings from a particular project researching the use of mathematics tasks and also from our previous research projects and experience in working with prospective and practising mathematics teachers. This chapter summarises the components of the project and outlines the foci of the chapters of the book as well as suggests how it might be read.

The Tasks Type and Mathematics Learning Project

The project, *Task Types in Mathematics Learning* (TTML), was a 3-year Australian Research Council funded research partnership between the Victorian Department of Education and Early Childhood Development, the Catholic Education Office (Melbourne), Monash University, and Australian Catholic University.

We worked with middle years' teachers of mathematics (grades 5–8) from three volunteer clusters of schools, one in each of the inner and outer suburbs of Melbourne and a nearly regional city, Geelong. Each cluster typically involved a secondary school and three or four primary schools. These three clusters represented a spread of socio-economic student backgrounds and included schools in both government and Catholic systems. Schools agreed to be involved in the research project and nominated particular teachers who would participate. As with most projects involving teachers and schools, not all teachers were enthusiastic participants, and the individuals' continuity of involvement varied over the course of the project, with teachers in secondary schools less likely to maintain a regular involvement, and the usual staff departures and new arrivals at schools.

P. Sullivan et al., *Teaching with Tasks for Effective Mathematics Learning*,
Mathematics Teacher Education 9, DOI 10.1007/978-1-4614-4681-1_1,

The project design involved each cluster of schools focusing on a particular type of task for an extended period of time (approximately 6 months), before moving on to a different task type. The three task types corresponded to those discussed in Chaps. 4, 5, and 6, respectively. Each of the principal researchers (Sullivan, Doug Clarke, and Barbara Clarke) took responsibility for one task type, moving from one school cluster to another after two school terms. In this way, each cluster of teachers received professional development in each of the task types, including the creation or sourcing of tasks related to the school's curriculum. Teacher professional learning programmes, a mixture of full-day and after-school sessions, focused on the nature of each task type, the associated pedagogies, ways of overcoming key constraints, and student assessment. For each task type, we set the teachers a goal of using at least one task of the relevant type in one lesson per week, with the expectation that teachers would increasingly generate their own tasks. Regular cluster meetings and annual conferences combining the three clusters allowed teachers opportunities to share experiences of teaching the tasks with schools from other clusters.

Once all teachers had experienced professional learning including classroom implementation of all three task types, our focus turned to the similarities and differences between the task types, their associated pedagogies, the constraints involved in using them, and the appropriate balance in the curriculum of the various task types.

As part of the project, we called for volunteers from schools to develop units of work (or lesson sequences) incorporating a range of task types. Groups of teachers from four schools agreed to do so. Supported by the research team, they developed units of work addressing volume and capacity, ratio and proportion, interpreting data, and financial literacy, respectively. As they taught these units, their classroom actions were studied closely by the research team, with field notes complemented by regular interviews with teachers before and after lessons. Teachers completed surveys on their experiences, as did students.

Our intention was to create optimal conditions for the successful implementation of each task type by ensuring that teachers had access to high-quality task exemplars and by supporting teachers on associated pedagogies. We also intended that the process of task creation and use was self-sustaining. One way we hoped to achieve this was through the development of a website on which we posted over 50 reports of lessons trialled by participating teachers, incorporating their experiences of teaching them.

Our research sought to explore factors that might influence or inhibit teachers' choice and use of such tasks. We used a model, shown in Fig. 1.1, adapted from Clark and Peterson (1986) in which there are four sets of background variables: teacher knowledge; teacher beliefs, attitudes, and self-goals; situational and other constraints; and teachers' intentions. The first three of these influence each other, and collectively they influence the fourth. A fifth set of variables, teacher actions and student outcomes, completes the model.

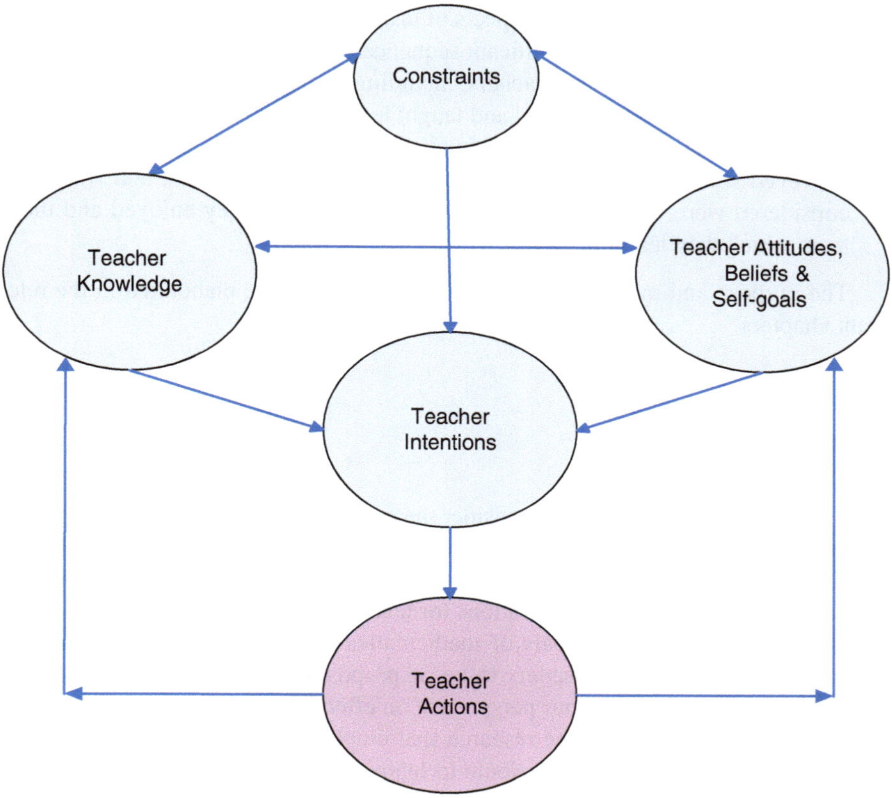

Fig. 1.1 A model of factors influencing task use

The goals of the TTML project were to describe the following:

- How different kinds of tasks respectively contribute to mathematics learning
- The features of successful exemplars of different types of tasks
- Constraints teachers might experience when using tasks and
- Teacher actions that can best support students' learning

The project had a number of components, many of which are elaborated in this book. We

- Surveyed teacher knowledge, and attitudes initially
- Held teacher professional learning sessions focusing on the nature of the various tasks; the associated pedagogies; ways of addressing key constraints, such as diversity in culture, language background, and readiness to learn; and student assessment
- Studied the implementation in classrooms of different mathematical tasks
- Surveyed teachers on their responses to the different tasks

- Surveyed students on various aspects of tasks and lessons
- Observed the teaching of significant sequences of lessons that were planned and taught by groups of project teachers, including interviewing the teachers
- Surveyed teachers who planned and taught lessons based on three different tasks designed to address the same mathematical content and
- Surveyed students on their reactions to different types of tasks, and what they considered were the types of mathematics lessons that they enjoyed and those from which they learnt the most

The purpose and methodology of the data collection are elaborated in the relevant chapters.

Using This Book

The intention is that readers will consider the contribution that well-designed mathematics tasks can make to student learning. In addition to reporting some key findings from the research, the book presents some examples of tasks as well as explains some of the key considerations for teachers when using these tasks.

Chapter 2 discusses the goals of mathematics teaching generally, it describes what mathematics teaching seems to be, it proposes what mathematics teaching could be, and it summarises our perspective on effective teaching.

Chapter 3 summarises some research that emphasises the importance of tasks, outlines the ways that tasks contribute to learning, and describes some key background variables underpinning teacher choice and use of tasks, including their knowledge, beliefs, constraints they experience, and their intentions.

Chapters 4–6 are similar in structure with each describing characteristics of particular categories of tasks: Chap. 4 describes tasks that address purposeful representations; Chap. 5 contextualised tasks; and Chap. 6 tasks that are open-ended. Each chapter explains the rationale for using such tasks, presents examples of the tasks, gives the details of a specific example and how it might be used, and summarises some research on teacher reactions to the tasks, respectively. Of course, these three types are not the only types of tasks that teachers use productively, but they each exemplify a particular approach, and it was these approaches that were the focus of the research.

Tasks are nearly always presented in the context of a classroom lesson. Chapter 7 describes the key decisions that teachers make in using a task to create lessons.

Lessons do not usually occur in isolation, but as part of a sequence of related experiences. Chapter 8 describes an aspect of the project in which teachers created lessons sequences and these sequences formed part of the data collection for the project.

Chapter 9 reports responses of students to data collection related to features of tasks and the extent to which they considered that particular types of tasks create the potential for enjoyment and learning.

Chapter 10 also reports student opinions, in this case their views on the characteristics of lessons that they enjoyed and learned from.

One aspect of the project was the creation of a set of three lessons that were based on the respective types of tasks. Chapter 11 reports responses of teachers and students to the three types of task.

Chapter 12 presents a summary of the key conclusions and recommendations arising from the research.

Chapter 13 presents five detailed examples of each type of task, partly so that the detail can illustrate the aspects of the tasks described throughout the book and partly to allow the respective characteristics of the tasks to be compared and contrasted.

The book is written to present the key information that may assist prospective and practising teachers to use tasks effectively in their teaching. Nevertheless, the chapters can generally be read separately. When information from a different chapter is needed to interpret the information given, this is noted, and readers can then readily refer to those chapters as well.

Chapter 2
Perspectives on Mathematics, Learning, and Teaching

This book is intended to provide some insights into the types of tasks that can assist students in learning mathematics. The insights though need to be interpreted in the context of our perspectives on some key issues. This chapter discusses what we see as the goals of teaching mathematics, it describes our perspectives on what it means to do mathematics, it presents a perspective on how students learn mathematics, and it lists some recommendations about effective mathematics teaching. The subsequent chapters use these perspectives as the basis of elaborations of the tasks that we discuss.

The Mathematics It Is Intended that Students Learn

Although sometimes seen as competing, there are two perspectives that inform the tasks that we present in this book.

The first perspective can be described as mathematical literacy or numeracy that emphasises the mathematics that students will need for work and for their lives generally. This perspective is represented in tasks that are connected to the experiences of students, that use contexts that are meaningful to them, or for which students can imagine the potential usefulness to them of learning to engage with the task. There is substantial numeracy in, for example, everyday experiences, in interpreting current events and priorities, in evaluating personal and civic priorities, in managing personal finances, in planning activities and projects, all of which all citizens are required to undertake for themselves. There is also considerable numeracy required in most workplaces, and school graduates who cannot cope with those numeracy demands have a restriction of work choices available to them (for more on this, see Bakker, Hoyles, Kent, & Noss, 2006; Human Capital Working Group, 2008; Zevenbergen & Zevenbergen, 2009).

The second perspective is mathematical. This perspective is represented in tasks that focus on definitions, principles, patterns, processes, and generalisations that have conventionally formed the basis of the mathematics curriculum, and which lay

P. Sullivan et al., *Teaching with Tasks for Effective Mathematics Learning*, Mathematics Teacher Education 9, DOI 10.1007/978-1-4614-4681-1_2,

the basis for much later school and university mathematics study. It incorporates appreciation of the elegance of mathematical thinking, and exposure to the wonder of topics like "pattern, symmetry, structure, proof, paradox, recursion, randomness, chaos, and infinity" (Ernest, 2010, p. 24).

We argue that these two perspectives are not competing and indeed are complementary. We argue that not only is it important for teachers to induct students into the discipline of mathematics, but it is also essential that students have opportunity to use the mathematics they are learning to solve practical problems of some relevance to them. The tasks in the following chapters seek to address both perspectives.

Fostering a Breadth of Mathematical Actions

One of the often repeated criticisms of most mathematics teaching is that it represents a narrow view of mathematics and hardly addresses the numeracy perspective at all (e.g., Hollingsworth, Lokan, & McCrae, 2003). One of the purposes of using a variety of task types is to offer students a breadth of perspectives on mathematics and numeracy and a variety of ways to learn. To allow a multidimensional perspective on what it means to do mathematics, we draw on the five strands of mathematical learning described by Kilpatrick, Swafford, and Findell (2001). Watson and Sullivan (2008) described these strands as follows:

> conceptual understanding—comprehension of mathematical concepts, operations, and relations;
> mathematical fluency—skill in carrying out procedures flexibly, accurately, efficiently, and appropriately, and, in addition to these procedures, factual knowledge and concepts that come to mind readily;
> strategic competence—ability to formulate, represent, and solve mathematical problems;
> adaptive reasoning—capacity for logical thought, reflection, explanation, and justification; and
> productive disposition—habitual inclination to see mathematics as sensible, useful, and worthwhile, coupled with a belief in diligence and one's own efficacy. (p. 112)

These five sets of mathematical actions contribute to providing a rationale for our emphasis on tasks. We use the term actions since these represent ways that students do mathematics. These actions have been incorporated into the Australian Curriculum: Mathematics [Australian Curriculum and Assessment Agency (ACARA, 2011)] and are described as proficiencies.

This way of thinking about mathematical actions has direct implications for the choice of tasks. For example, if a teacher is keen to develop procedural or mathematical fluency, then this can be achieved by providing representative worked examples followed by repetitious practice. Of course, fluency is indeed a critical aspect of learning mathematics, but we do not focus on tasks that develop fluency in this book since such tasks are well represented in every school mathematics text we have seen.

If a teacher seeks to develop conceptual understanding, then it is possible to do this by using clear and interactive explanations, by encouraging communication between teacher and students and between students. This though can also be facilitated by choosing tasks that exemplify the underlying aspects of mathematics, such as in tasks that incorporate the use of models and representations. The tasks described and elaborated in Chap. 4 are examples of such tasks.

We see strategic competence and adaptive reasoning as key actions for students when learning and doing mathematics and argue that it is not possible to engage students in such actions through explanations and worked examples. It is necessary for students to work on tasks which require them, for example, to make and justify choices, to integrate ideas, to plan strategies, and to explain their thinking. Such tasks are the focus of Chaps. 5 and 6 in this book.

Considering Students' Perspectives on Mathematical Tasks

A further set of assumptions that underpin our approach to task choice and use is the ways that students might approach tasks.

One aspect is a perspective on what constitutes knowledge. We accept the social constructivist view, as summarised by Ernest (1994), that recognises knowing as active, "individual and personal, and that it is based on previously constructed knowledge" (p. 2). In this, knowledge is not fixed, rather it is socially negotiated, and is sought and expressed through language. In other words, students do not learn only by listening but also by engaging in experiences that contribute to their learning, and having the opportunity to share approaches to tasks with others.

A related aspect is the level at which tasks are pitched. A metaphor that is helpful for this is Vygotsky's (1978) zone of proximal development (ZPD) which he described as the "distance between the actual developmental level as determined by independent problem solving and the level of potential development as determined by problem solving under adult guidance or in collaboration with more capable peers" (p. 86). In other words, tasks should go beyond the current level of students' knowledge, but be within that range within which students can reasonably be expected to engage with support.

A further aspect is the extent to which students choose to engage with the experiences associated with working on a task. Fredericks, Blumfield, and Paris (2004) categorise engagement as behavioural, affective, and cognitive. We are interested in tasks that promote cognitive engagement, but of course affective engagement is relevant in that students must see some point in choosing to work on a task. Connected to this are the key characteristics of motivation, described by Middleton (1995) as interest, control, and arousal. Another relevant perspective was proposed by Dweck (2000) who categorized students as either having a performance or mastery orientation. Students with a performance orientation seek social affirmation as the goal of their effort rather than understanding of the content, and avoid risk taking and challenging tasks due to fear of failure. In contrast, students with a mastery orientation seek to understand the content, and evaluate their success by whether they feel they can use and transfer their knowledge. They remain focused on mastering skills and knowledge even when challenged, they do not see failure as an indictment on themselves, and they believe that effort leads to success. In other words, it seems helpful if tasks allow students opportunity to have a sense of control by allowing them to make decisions, are interesting to the students, incorporate a rationale for them to engage, provide some challenge, reduce the risk of failure, and for which success provides the motivation for further engagement. The tasks described in Chaps. 4–6 are proposed as examples of such tasks.

Approaches to Teaching Mathematics

Our approach to task choice and use is connected to the pedagogies that are used to support the student engagement with the task. There are two sets of pedagogical approaches presented here that are then referred to in various places throughout the book.

The first set of pedagogical approaches was developed by Doug Clarke and Barbara Clarke (2004), arising from detailed case studies of teachers who had been identified as particularly effective in the Australian Early Numeracy Research Project. Their list is grouped under ten headings and 25 specific pedagogical actions. While their list was drawn from research with early years' mathematics teachers, we argue that the headings and actions are applicable at all levels. Their list is written in the form of advice to teachers:

Mathematical focus	Focus on important mathematical ideas
	Make the mathematical focus clear to the children
Features of tasks	Structure purposeful tasks that enable different possibilities, strategies, and products to emerge
	Choose tasks that engage children and maintain involvement
Materials, tools, and representations	Use a range of materials/representations/contexts for the same concept
Adaptations/ connections/links	Use teachable moments as they occur
	Make connections to mathematical ideas from previous lessons or experiences
Organisational style(s), teaching approaches	Engage and focus children's mathematical thinking through an introductory, whole group activity
	Choose from a variety of individual and group structures and teacher roles within the major part of the lesson
Learning community and classroom interaction	Use a range of question types to probe and challenge children's thinking and reasoning
	Hold back from telling children everything
	Encourage children to explain their mathematical thinking/ideas
	Encourage children to listen and evaluate others' mathematical thinking/ideas, and help with methods and understanding
	Listen attentively to individual children
	Build on children's mathematical ideas and strategies
Expectations	Have high but realistic mathematical expectations of all children
	Promote and value effort, persistence, and concentration

(continued)

(continued)

Reflection	Draw out key mathematical ideas during and/or towards the end of the lesson
	After the lesson, reflect on children's responses and learning, together with activities and lesson content
Assessment methods	Collect data by observation and/or listening to children, taking notes as appropriate
	Use a variety of assessment methods
	Modify planning as a result of assessment
Personal attributes of the teacher	Believe that mathematics learning can and should be enjoyable
	Are confident in their own knowledge of mathematics at the level they are teaching
	Show pride and pleasure in individuals' success

A second set of pedagogical actions associated with mathematics teaching was developed as a prompt to school-based collaborative teacher learning, drawing on the Clarke and Clarke actions and similar recommendations such as Hattie and Timperley (2007) and Education Queensland (2010). The following presents six key principles of effective mathematics teaching. Sullivan (2011) had been working with teachers who had been presented with quite specific recommendations of ways to teach reading and literacy, and who had requested a similar specific list for mathematics teaching. These principles, also used in various discussions through this book to describe teacher thinking and actions, also written in the form of advice to teachers, are as follows:

Principle 1. Identify big ideas that underpin the concepts you are seeking to teach, and communicate to students that these are the goals of the teaching, including explaining how you hope they will learn.

Principle 2. Build on what the students know, mathematically and experientially, including creating and connecting students with stories that both contextualise and establish a rationale for the learning.

Principle 3. Engage students by utilising a variety of rich and challenging tasks that allow students time and opportunities to make decisions and which use a variety of forms of representation.

Principle 4. Interact with students while they engage in the experiences; encourage students to interact with each other including asking and answering questions, and specifically planning to support students who need it; and challenge those who are ready.

Principle 5. Adopt pedagogies that foster communication and mutual responsibilities by encouraging students to work in small groups, and using reporting to the class by students as a learning opportunity.

Principle 6. Fluency is important, and it can be developed in two ways: by short everyday practice of mental calculation or number manipulation; and by practice, reinforcement and prompting transfer of learnt skills.

These principles together with aspects from the above lists are used in interpreting some of the research results and recommendations for practice in the following chapters. Note that central to both these lists is the role and contribution of tasks.

Summary

The approaches to mathematics teaching described in this book are based on consideration of the goals of mathematics teaching, current approaches to teaching mathematics taken by many teachers, and recommendations about what actions seem likely to create experiences that are accessible by all students and which engage students in learning and creating mathematics.

We argue that an important component of understanding teaching and improving learning is to identify the types of tasks that prompt engagement, thinking, the making of cognitive connections, and the associated teacher actions that support the use of such tasks, including addressing the needs of individual learners. The challenge for mathematics teachers is to foster mathematical learning, and the key media for pedagogical interaction between teacher and students are the tasks in which the students engage.

The underlying argument, then, is that mathematical learning experiences are based on tasks; that the better the task (appropriately used), the better the opportunities for effective teaching; and the better the task, the better the learning.

Chapter 3
Tasks and Mathematics Learning

This chapter summarises some research results and scholarly commentary that emphasise the importance of mathematical tasks, and the ways they contribute to learning, the role of teacher knowledge in the effective use of tasks, the ways that teacher beliefs and attitudes influence the use of tasks, the nature of the constraints that teachers can anticipate, and the ways that each of these influence the teacher intentions. Particular recommendations about task creation, selection, and use are presented.

Introduction

We argue that more than any other actions that teachers might take, posing tasks that engage students in thinking for themselves about mathematics is the main stimulus for student learning (Anthony & Walshaw, 2009). Drawing on Watson and Sullivan (2008), we use the term *task* to refer to information that serves as the prompt for student work, presented to them as questions, situations, and instructions that are both the starting point and context for their learning. We use the term *activity* to refer to the thoughts and actions, physical, spoken, written, and recorded, that students take in response to the task.

A framework that can guide thinking about tasks was proposed by Stein, Grover, and Henningsen (1996) who argued that the consideration of classroom tasks by teachers goes from the following:

> Mathematical task as presented in instructional materials
> which, influenced by the teacher goals, their subject matter knowledge, and their knowledge of students, informs …
> … mathematical task as set up by the teacher in the classroom
> which, influenced by classroom norms, task conditions, teacher instructional habits and dispositions, and students learning habits and dispositions, influences …
> … mathematical task as implemented by students
> which creates the potential for …
> … students' learning.

P. Sullivan et al., *Teaching with Tasks for Effective Mathematics Learning*,
Mathematics Teacher Education 9, DOI 10.1007/978-1-4614-4681-1_3,

The Connection Between Tasks and Learning

Our research and this book are based on an assumption that choice of tasks and the associated pedagogies are key aspects of teaching and learning mathematics. There is substantial support for this view. Anthony and Walshaw (2009), for example, in a synthesis of research results that inform practice, argued that the role of mathematical tasks, activities, and tools is central. They concluded that "in the mathematics classroom, it is through tasks, more than in any other way, that opportunities to learn are made available to the students" (p. 96). Likewise the National Council of Teachers of Mathematics (2000) had task choice and use as one of the principles of effective teaching. Christiansen and Walther (1986) explained that the tasks given to students, and their associated activity, serve as the interaction between the teacher and the learner. They argued that non-routine tasks, because of the interplay between different aspects of learning, provide optimal conditions for cognitive development in which new knowledge is constructed relationally and items of earlier knowledge are recognised and evaluated. It follows that the best tasks are those that provide appropriate contexts and complexity; that stimulate construction of cognitive networks, thinking, creativity, and reflection; and that address significant mathematical topics explicitly. Tasks are clearly important.

An underlying premise to our approach to mathematics teaching is that engagement in mathematical thinking comes from students working on a succession of problem-like tasks, rather than following the teacher's instructions step by step. Cobb and McClain argued that teachers should have a clear impression of the direction that the learning of the individuals and the class will take. They proposed that the teacher should form an "instructional sequence (that) takes the form of a conjectured learning trajectory that culminates with the mathematical ideas that constitute our overall instructional intent" (Cobb & McClain, 1999, p. 24). The learning occurs as a product of students working on tasks, purposefully selected by the teacher, that form the basis of an ongoing dialogue with the teacher and with peers on their strategies and products.

Various comparative studies have endorsed the focus on rich and challenging tasks. Hiebert and Wearne (1997), for example, analysed teaching approaches and tasks in six classrooms. Four of the teachers used a conventional textbook approach. Two teachers used an alternative approach characterised by having fewer tasks but spending more time exploring and discussing them, having fewer pencil-and-paper based tasks and posing more problem situations, and asking students more higher-order questions. At the end of the year, students in the classrooms using this alternate approach gave longer responses and demonstrated higher levels of performance on mathematical assessments. Likewise, Stein and Lane (1996) reported that the greatest gains on performance assessments, including questions that required high levels of mathematical thinking and reasoning, were related to the use of instructional tasks that engaged students in "doing mathematics or using procedures with connection to meaning" (p. 50). They further noted that student performance gains were greater when "tasks were both set up and implemented to encourage use of multiple solution strategies, multiple representation and explanations" (p. 50).

We also argue that what students learn is determined largely by the tasks they are given. Tasks designed to prompt higher-order thinking are more likely to produce such thinking than tasks designed to offer skills practice (see e.g. Doyle, 1986; Hiebert & Wearne, 1997).

Types of Tasks for Mathematics Teaching

Because tasks are central to teaching, various authors have synthesised examples of types of mathematical tasks. Anthony and Walshaw (2009) for example, identified a range of tasks that have been reported and analysed in the literature, including those that pose problems for students to solve and have a mathematical focus; those that require students to model thinking and promote reflection; those that prompt discussion of aspects that vary; those that do not, as well as those that ask students to interpret and critique data with scepticism, and those that prompt sense making and justification of thinking. Anthony and Walshaw (2009) argued that, above all, tasks should have a mathematical focus, should be challenging for the learner, and provoke insights into the structure of mathematics and strategies or methods for solving problems. They argued that "whatever their format, effective tasks are those that afford opportunities for students to investigate mathematical structure, to generalise, and to exemplify" (p. 141). Swan (2005) presented carefully designed tasks using particular categories that he described as classifying mathematical objects, interpreting multiple representations, evaluating mathematical statements, creating problems, and analysing reasoning and solutions.

In other words, there is a range of types of tasks that teachers can use to support students' learning of mathematics. This book presents three types of task that can support mathematical teaching that we describe as representational, contextual, and open-ended. These are elaborated respectively in Chaps. 4, 5, and 6.

The Role of Teacher Knowledge in Effective Task Use

The ways that teachers choose and use tasks are connected to what they know. There is now substantial evidence (e.g. Rowland, Huckstep, & Thwaites, 2005) that teachers' knowledge of mathematics influences their effectiveness in facilitating learning. This is especially the case when using tasks, explanations, and illustrations that teachers prepare in order to support student learning require understanding of the mathematics. Their capacity to interpret and react to students' mathematical responses is also important (Fennema & Franke, 1992).

There are two major categories of knowledge needed to convert tasks to lessons, and for the teaching of mathematics generally: subject matter knowledge and pedagogical content knowledge. Hill, Ball, and Schilling (2008) described diagrammatically components of these two types of knowledge.

The first category, subject matter knowledge, is made up of the sub-categories of common content knowledge, specialised content knowledge, and knowledge at the mathematical horizon. Common content knowledge is the mathematics needed to solve a task. This is the type of knowledge that someone who is not a teacher but who is mathematically strong might have. Specialised content knowledge is

> the knowledge that allows teachers to engage in particularly teaching tasks, including how to accurately represent mathematical ideas, provide mathematical explanations for common rules and procedures, and examine and understand unusual solution methods to problems. (Hill et al., 2008, p. 378)

This includes recognising that a task can be solved in different ways, that there is value in considering alternate solution paths, and that there are particular strategies that can be applied to solving unfamiliar problems. *Specialised content knowledge* is described as the mathematical knowledge and skills uniquely needed by teachers in the conduct of their work. In looking for patterns in errors made by students or in considering whether a student-generated solution strategy could be generalised to other tasks, a teacher is drawing upon this knowledge.

Knowledge at the mathematical horizon is less clearly defined, but we take it to mean knowing where the current content will be used in future learning, and the ways that the current topic connects to the mathematics to be learned in subsequent years.

Clearly all three of these categories of knowledge are important for effective task use, but most important of all is the specialized knowledge needed by teachers.

The second of the Hill et al. (2008) categories, pedagogical content knowledge (PCK) is made up of the sub-categories of knowledge of content and teaching, knowledge of content and students, and knowledge of curriculum. PCK had been earlier described by Shulman (1986) as

> an understanding of what makes the learning of specific topics easy or difficult: the conceptions and preconceptions that students of different ages and backgrounds bring with them to the learning of those most frequently taught topics and lessons. If those preconceptions are misconceptions, which they so often are, teachers need knowledge of the strategies most likely to be fruitful in reorganizing the understanding of learners. (pp. 9–10)

Hill et al. delineated three sub-categories of PCK. The first, *knowledge of content and teaching*, includes an understanding of how to sequence particular content for instruction, of how to evaluate instructional advantages and disadvantages of particular representations, and of the knowledge required to make "instructional decisions about which student contributions to pursue and which to ignore or save for a later time" (p. 401).

The second sub-category, *knowledge of content and students*, is "knowledge that combines knowing about students and knowing about mathematics" (p. 401). This includes, for example, anticipating students' cognitive and affective responses to particular tasks and what they will find easy and hard. Teachers "must also be able to hear and interpret students' emerging and incomplete thinking as expressed in the ways that pupils use language" (p. 401).

The third sub-category, *knowledge of curriculum*, is slightly different in that it relates closely to Shulman's (1986) "curricular knowledge", in which curriculum "is represented by the full range of programs designed for the teaching of particular subjects and topics at a given level, the variety of instructional materials available in relation to those programs, and the set of characteristics that serve as both the indications and contraindications for the use of particular curriculum or program materials in particular circumstances" (p. 10).

One of the early phases of the project was to examine teacher knowledge. We used the Hill et al. categorisation to examine responses of teachers of students aged around 10–14 years to a range of items. One of these items asked them to create a lesson out of a task requiring students to decide which is bigger: 2/3 or 201/301 (see Sullivan, Clarke, & Clarke, 2009) for a full discussion of this). Consideration of responses to this task can elaborate the categories of teacher knowledge discussed above.

Unless a teacher is planning to just "throw" the fraction comparison task to the class without having solved it personally, we believe that it is clear that a teacher needs relevant *common content knowledge* to determine which of 2/3 and 201/301 is larger for his or herself. Our observations during the completion of the questionnaire indicated that this was problematic for a number of teachers. Given our belief that an understanding of relevant *common content knowledge* is a necessary condition for successful translation of such tasks into classroom use, this may explain the inadequacies of a number of the responses of teachers in this survey.

Conversations with teachers around this task indicated that many were not aware of the wide variety of strategies that students might bring to it, aside from attempting to find common denominators or using a calculator to divide the respective numerators by their denominators. Such awareness draws upon *specialised content knowledge*, which we see as critical for effective teaching.

Although it could be argued that the content is not appropriate for the students of those grade 5 and 6 teachers who completed the survey, one would nevertheless hope that the teachers possessed sufficient *knowledge at the mathematical horizon* to discuss how the problem might be used in a junior secondary classroom.

We are also somewhat concerned that many teachers seemed unable to describe the mathematical content in terms which would have indicated that they realised this task was actually about fraction comparison. This connects directly to *knowledge of content and curriculum*. The responses call into question the sense teachers make of curriculum documents including syllabuses (i.e. the intended curriculum), when this is a difficulty.

It is possible that *knowledge of content and students* had the effect of limiting the vision that teachers had for the use of the task.

It is clear that many teachers found translating the fraction comparison task into a worthwhile learning experience for middle school students difficult, or at least had great difficulty in articulating how they might do so. It was interesting that primary and secondary teachers were equally likely to create student-focused investigative type lessons, which is contrary to some common conceptions of differences between such teachers.

It became clear to the project team that, particularly when considering topics that both students and teachers find difficult (such as rational number), professional development leaders need to take the time to focus on all six components of knowledge for teaching mathematics. It is also clear that we need to give a greater time commitment to discussing how such content might be addressed across the middle years. Importantly, we should not take for granted that all or even most teachers can necessarily translate a good idea or task into a worthwhile learning experience for students, without considerable professional development support.

The knowledge of teachers clearly influences their choices of tasks and the way they use them, but this is quite a specific form of knowledge. It is unlikely that additional studies of mathematics after graduation are likely to improve task choice and use, but specific attention to those aspects of specialised content knowledge that can support effective task choice and use seems critical.

An interesting aspect of this knowledge is that teachers cannot be expected to know all the mathematics they will need in their teaching. They do however need the skills and orientation to finding out about the necessary mathematics when they need it. Perhaps teachers also need the willingness to say to their students "I do not know the answer, but let us see how we can find this out".

Teacher Beliefs, Attitudes, and Self-Goals

Another set of variables influencing implementation of tasks relates to teacher beliefs, attitudes, and self-goals, particularly their beliefs about mathematics and learning mathematics (Thompson, 1992), and their attitudes to mathematics generally (Hannula, 2004). These beliefs are manifest in the kinds of tasks posed and the forms of affirmation that teachers use, as is the degree of teachers' self-efficacy (Tschannen-Moran, Hoy, & Hoy, 1998), which is the extent to which they believe they can influence students' learning despite contextual constraints.

Philipp (2007) noted that terms such as beliefs, attitudes, and conceptions are not used in the literature in a consistent way, and the relationships between these are not generally agreed upon. However, his working definitions are helpful for our purpose here. He claimed that "beliefs are psychologically held understandings, premises or propositions about the world that are thought to be true. Beliefs are more cognitive, are felt less intensely, and are harder to change than attitudes", while conceptions were seen as a broader term "encompassing beliefs, meanings, concepts, propositions, rules, mental images and preferences" (p. 259).

In her seminal work reviewing the literature on beliefs and conceptions, Thompson (1992) focused upon teachers' conceptions of the nature of mathematics, their conceptions of mathematics teaching and learning, and, importantly for this chapter, the relationship between these conceptions and instructional practice.

In relation to teachers' conceptions of the nature of mathematics, there have been several categorisations, including three of Ernest (1988), which can be summarised as (a) a dynamic, problem-driven discipline; (b) a static, unified body of knowledge; and (c) mechanistic, meaning a bag of tools. In applying this three-part categorisation, Nisbet and Warren (2000) used a 56-item Likert-scale survey to gather data on 398 primary teachers' views on teaching, learning, and assessing mathematics, which were analysed using factor analysis. Interestingly, only the static and mechanistic views of Ernest above emerged as factors.

Similarly, Kuhs and Ball (1986) proposed four views of the teachers' conceptions of teaching and learning: learner focused, content focused with an emphasis on conceptual understanding, content focused with an emphasis on performance, and classroom focused.

An interesting aspect of this is that Thompson (1992) noted a stronger relationship found by researchers between teachers' conceptions of mathematics and their instructional practice than between their conceptions of teaching and learning and their instructional practice. Such discrepancies highlight a methodological concern in relation to how beliefs or conceptions are measured (Philipp, 2007). In fact, Thompson recommended that researchers move beyond professed beliefs and "examine teachers' verbal data along with observational data of their instructional practice or mathematical behaviour" (p. 135).

It is clear that the relationships between beliefs, attitudes, conceptions, and practices are complex, and classroom researchers need to take into account not only the individuals variables but also the interaction between them.

Constraints

A further set of variables relate to the constraints teachers might experience in the implementation of tasks. One major constraint can be diversity in students' cultural background, language fluency, and readiness to learn (see Delpit, 1988). Other constraints include that the structure of classrooms and common responses of students to schooling restrict teacher choices (see Doyle, 1986), and the skill levels of the students inhibit their participation in non-routine tasks, especially in the middle years (see Stein & Lane, 1996). This latter constraint is particularly critical, and connects to the challenge teachers experience due to the diversity of readiness they experience in the classes they teach.

Our research examined ways that teachers can support students who experience difficulty with a task that is set for the class. It is common in some classrooms for teachers to gather students experiencing difficulty together and teach them as a group, probably more slowly, with more repetition of procedural steps. We suggest that students are more likely to feel fully part of the class, and so be more likely to choose to participate fully, if teachers offer prompts that allow those students experiencing difficulty to engage in active experiences related to the initial goal task, rather than, for example, requiring such students to listen to additional explanations or assuming that they will pursue goals substantially different from the rest of the class. Sullivan, Mousley, and Zevenbergen (2006) termed these *enabling prompts* and suggested that effective prompts can include

- Reducing the required number of steps
- Simplifying the modes of representing results
- Making the task more concrete or
- Reducing the size of the numbers involved

ZPD also provides a metaphor for the support that teachers can offer to students experiencing difficulty. If, for example, the teacher poses problems that are challenges for all students, in most classes there will be some students who are not already at the level of independent problem solving for this particular problem. We argue that adult guidance or peer collaboration might be offered to such students through adapting the task on which they are working, as distinct from, for example, grouping students together and having a group undertake quite different work.

This notion of adapting tasks is a recurring theme in advice to teachers (e.g. the Association of Teacher of Mathematics, 1988; Griffin & Case, 1997). Anthony and Walshaw (2009) likewise emphasised the importance of teacher considerations of the needs of diverse learners; connecting tasks to learning goals, to the learners' existing proficiencies and knowledge; and the role of context. This complements suggestions from Gee (2004) that tasks be able to be customised to match the readiness of the learner, both for those who experience difficulty and for those for whom the core task is not challenging. It is stressed that adaptation of tasks is not intended to remove challenge, but to allow students to engage in a preliminary task. Examples of such prompts are presented in each of the following three chapters.

A further aspect relates to teachers anticipating that some students may complete the planned tasks quickly and posing supplementary tasks that extend their thinking on that same task, rather than proceeding onto content planned for the next lesson. Sullivan et al. termed these *extending prompts*.

Overall, the aim is to move the learners forward so that every student has had experiences that allow them to engage in collective mathematical argumentation, reflection, and dialogue and that all will be ready for the subsequent lesson. The linguistic, psychological, and cultural nature of classroom talk contributes to the development of deeper communal understanding of mathematical concepts and principles (Brown & Renshaw, 2006), and our hope is that all students can participate in the classroom task.

Constraints experienced by teachers in the project are discussed in each of the next three chapters.

Teacher Intentions

Another key component of the model is teachers' intentions or, in other words, how they plan to teach. Each of the above factors influences teachers' decisions about what they plan to do in their lessons. As the model indicates, teachers decide what they hope to achieve, how they might achieve it, what might inhibit this, and how they will overcome barriers.

One of the interesting results from Stein et al. (1996) was the tendency of teachers to reduce the level of potential demand of tasks. Doyle (1986) and Desforges and Cockburn (1987) attribute this phenomenon to complicity between teacher and students to reduce their risk of making errors. Tzur (2008) argued that there are substantial deviations between ways that developers intend tasks to be used and actions that teachers take. Tzur suggested that there are two key ways that teachers modify tasks: at the planning stage if they anticipate that the task cannot accomplish their goals; and once they see student responses if they are not as intended. Charalambous (2008) argued that the mathematical knowledge of teachers is one factor determining whether they reduce the mathematical demand of tasks based on their expectations for the students.

It seems that it is possible to assist teachers to be aware of this tendency. This was also a central feature of the teacher learning sessions that we conducted with our project teachers, and we focussed on ways that teachers can maintain their intended level of challenge in their teaching.

Summary

In summary, it seems that the choice and use of tasks are central to effective mathematics teaching, and that the nature of learning is connected to characteristics of the task. Some of the key characteristics of tasks that authors recommended are that they

- Engage students in doing important mathematics, fostering meaning making, understanding, and connections to other aspects of mathematics
- Are challenging for most of the class, with the pathway to the solution not being obvious to the students
- Require students to think, make decisions, and communicate with each other
- Prompt thinking and reflection and,
- Use contexts or situations with which the students are familiar and which they see as potentially useful for them or connected to their lives

Of course, there are few tasks that can do all of these things, but the challenge for the teacher is to use these ingredients to create a balanced and healthy diet, even though the individual "meals" may include only a few of these ingredients.

The chapter also elaborated key background variables including teachers' knowledge, their beliefs, and the constraints they experience. These background variables are elaborated further in subsequent chapters, especially the way the tasks can be adapted to cater for the diversity of student readiness, and the match between teacher intentions and their actions.

Chapter 4
Using Purposeful Representational Tasks

This is the first of three chapters that elaborate the particular categories or types of tasks that were the focus of our project. In considering the range of tasks that can be effectively used to support student mathematics learning, tasks that make use of materials and visual support need to be considered. In this chapter, we consider tasks that use models, tools, and representations, present some examples, and discuss the perceived benefits and challenges for teachers in their use. We term them purposeful representational tasks.

A Rationale for Purposeful Representational Tasks

The use of models, tools, and representations is a key component of effective mathematics teaching (Clarke & Clarke, 2004). Appropriate models and representations, in the hands of capable teachers, support children's conceptual development and can build skills. Tasks involving models and representations are associated with good traditional mathematics teaching (see Watson & Mason, 1998), but their use is not always evident in the regular classroom.

In considering a range of task types, we were committed to including those tasks that use models or representations to support student learning. Such tasks enable the foregrounding of the mathematics through use of materials providing opportunities for students' cognitive engagement. The mathematics in the task is made explicit to the students. Explicit is used here to emphasise that the mathematics was made explicit in the use of the task, not to imply that the teacher was "telling" the student

P. Sullivan et al., *Teaching with Tasks for Effective Mathematics Learning*, Mathematics Teacher Education 9, DOI 10.1007/978-1-4614-4681-1_4,

what to do, without appropriate student decision making. We were not looking for tasks that reproduce the type of highly structured mathematics lesson which typically focuses on developing procedural competence, but rather for conceptually focused tasks which purposefully make use of models, representations, or other tools—"good" tasks that engage and challenge students and provide opportunities for developing understanding.

The mathematical purpose is pivotal in these tasks, and as a result, it influences teacher decision making. Arguably, there is likely to be less opportunity for a detour from the plan than with other more open types of task, but also the teacher may be less willing to deviate into a different area of mathematics, when faced with an opportunity, given the clear mathematical intent behind the choice of task.

The identification of quality tasks of this type was challenging for us and, as is outlined later, was challenging for teachers. Clearly there is a category of tasks that makes use of a model or visual tool to represent the mathematics and support student learning. We acknowledge that the use of concrete or manipulative materials is relatively common practice in junior primary classrooms but is less evident in the middle years (ages 10–14 years).

Studies have shown that using manipulative materials is successful for developing students' understanding of concepts such as area (Cass, Cates, & Smith, 2003) and fractions (Butler, Miller, Crehan, Babbitt, & Pierce, 2003), some of the key topics in the middle years. However, these need to connect directly to the relevant mathematical understandings to be effective (Ball, 1992). Boulton-Lewis and Halford (1992) acknowledged the different ways in which manipulatives can be interpreted in that "some of these are concrete embodiments of mathematical concepts and processes and others are representations inherent in the discipline of mathematics" (p. 1). Stacey, Helme, Archer, and Condon (2001) also suggested that the transparency of the materials to illustrate clearly the intended mathematical concept was critical to their successful use.

In addition, there was a category of task that focuses on aspects of mathematics that can be represented in different forms (such as rational number—fractions, percentage, and decimals), and learning is supported through matching or comparing. They involved the manipulation of specifically designed materials, but it was the comparison that facilitated the learning. Swan (2006) argued strongly for the value of this type of task in supporting the learning of mathematical concepts and the need to make these challenging rather than simply a revision or consolidation task.

It is not the intention here to provide a detailed categorisation of different types of tasks but to acknowledge that a range of purposeful representational tasks were discussed in the development of the project and were also shared with the teachers as exemplars.

Defining Purposeful Representational Tasks in the TTML Project

In describing these tasks, we are referring to focused experiences that engage children in developing and consolidating mathematical understanding, using models and representations as a basis of this. An example is a teacher who uses a fraction wall to provide a linear model of fractions, and poses tasks that require students to compare fractions, to determine equivalences, and to solve problems involving fractional operations. The fraction wall simplifies the mathematical complexity (see Colour in Fractions example below).

In the initial TTML professional development, teachers were provided with the following definition of this type of task:

> The teacher commences with an important mathematical idea, and proposes tasks that involve models or representations or tools, which help students to understand the mathematics. There is no attempt to link mathematics to its practical applications. For example, the use of a fraction wall in a chance game can assist in developing an understanding of equivalence, improper fractions, and simple operations with fractions.

With a task of this kind, the mathematics is inherent in the model, representation, or materials provided, and limited teacher explanation is required before the task is commenced by students. An introduction is given by the teacher with a focus on tuning into the mathematics and reviewing some prerequisites. The following descriptions elaborate this type of task.

An Example of a Purposeful Representational Task: Colour in Fractions

Purposeful representational tasks often involve a game to engage students, but always with a clear mathematical focus. Colour in Fractions (see Doug Clarke & Roche, 2010 for detailed discussion of the implementation of this task) uses a representation (the fraction wall) in a game context. The set-up and rules are given below.

Colour in Fractions

Students have two dice that when thrown and combined as a pair, create fractions up to twelfths, and a fraction wall. They colour in sections of the wall that correspond to the fractions that they roll with the dice.

- One die labelled 1, 2, 2, 3, 3, 4 in one colour (the numerator)
- Another die labelled */2, */3, */4, */6, */8, */12 in another colour (the denominator)

The wall is like this:

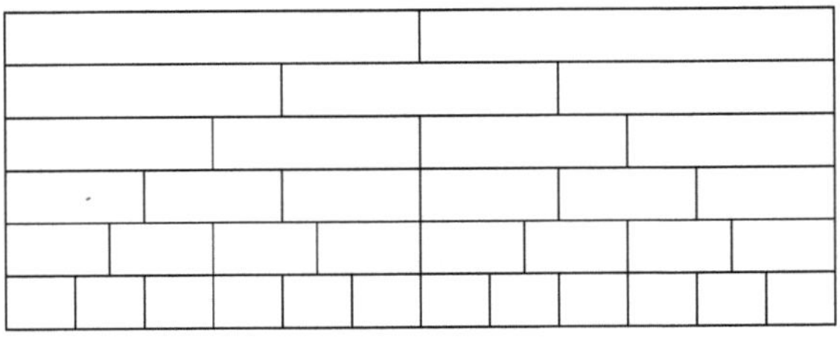

Players in turn throw both dice. They make a fraction, the first die being the numerator. Each horizontal strip across the wall is one whole.

They then colour the equivalent of the fraction shown. For example, if they throw 2 and *⁄4, then they can colour in 2/4 of one line or 4/8 of one line or 1/4 of one line and 2/8 of another, or any other combination that is the same as 2/4.

If a player is unable to use his/her turn, he/she "passes". The first player who colours in the whole wall is the winner.

There can be further mathematical questions posed based on the game and class discussion of strategies.

Using this game, the teacher is likely to have a clear mathematical intent on developing students' understanding of equivalence, improper fractions, and addition of fractions, with the "fraction wall" model as a key component. That being said, different students may draw quite different learning from the experience of playing the game. One student may discover for the first time that 2/3 is equivalent to 8/12 and why, while another may determine that when left with 1/12 to fill at the end of the game, there is only one chance in 36 of a successful roll each time.

The importance of the model and the explicit focus on the mathematics are features of this task, and the fraction wall has the potential to provide an on-going tool for the student to use in other situations. It is consistent with the advice provided to the project teachers through systemic guidelines:

> The aim of learning with a model is to give a student a tool to think with; something that they can draw upon to interpret symbolic work. So models should be carefully chosen and used thoroughly and consistently for some time. (Mathematics Continuum, DEECD, 2007)

In addition to model-based tasks, sorting and classifying activities, as discussed earlier, are considered to be purposeful representational tasks. These can take the form of different representations of the same mathematics such as common fractions, decimals and percentages, or graphical representations of data that are matched. Such activities can also involve more open processes such as the sorting of algebraic equations to identify and highlight the common features where the students are required to provide their criteria to the teacher.

One further category of task also involved carefully designed materials. These created an opportunity for students to explore a mathematically rich and often complex relationship. For example using percentages to increase quantities from the work of Swan involved the following arrangement of four money cards (adapted for Australian currency):

The students were then provided with arrow cards and asked to arrange them in order to indicate the relationship between the different values of the money card. For example the card "up by 20%" would hopefully be placed in between $100 and $120 with the arrow pointing towards $120. This was a challenging task focusing on percentage increase and decrease. Additional sets of cards were provided that express the relationships in terms of fractional and decimal increase (e.g. up by one quarter or × 0.5). It is possible to extend the experience by inviting the students to create their own set.

While purposeful representational tasks are usually not contextualised, there is sometimes a "hook" that helps to engage the students. One example is the *Chocolate Block* task (see Chap. 13 for an outline of the task and Doug Clarke (2006) for more details on the implementation and justification), where sharing of chocolate represents not only an engaging context but also a model for the development of the concept of fraction as division.

We also provided the teachers with specific advice on the pedagogies associated with this type of task. In particular, we suggested the following:

> These tasks involve the introduction to, or use of models, representations, tools, or explanations that elaborate or exemplify the mathematics. Following student work on the task, the teacher leads a discussion on the mathematics that has emerged from the task, and will seek to draw out commonalities and generalisations.

Some Additional Examples

Following are some brief descriptions of additional examples that illustrate further the nature of these tasks:

- *Fractions close to 1.* Place number cards (choosing from 1, 3, 4, 5, 6, 7) in the boxes to make fractions so that when you add them, the answer is as close to one as possible, but not equal to one. Each card can be used only once.

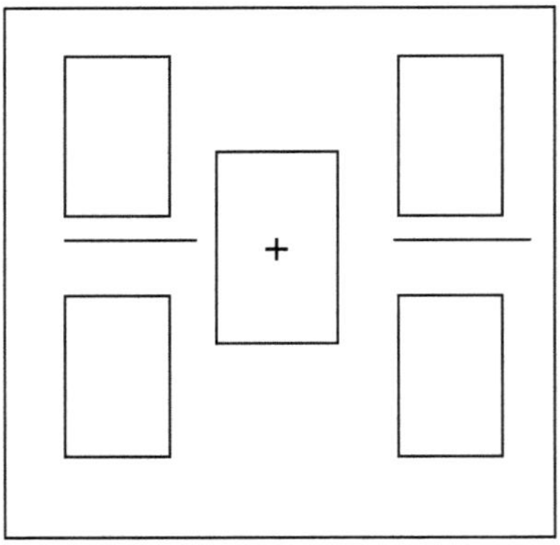

- *Sorting Equations.* Students are presented with a range of equations in various forms, and asked to sort them into two groups in a way which makes sense to them, justifying their sort *(e.g.* $x + 7 = 23$; $3x + 2 = 5x - 8$; $x/7 = 3$; $6 - x = x + 11$; $3.3x + 8 = 17$*).*
- *Estimation with Fractions.* Students estimate a point two-fifths of the way across the board and then discuss how to determine the actual point. A measurement model is used and then extended to further examples (Lovitt & Clarke, 1988).
- *Fraction, Decimal, and Percentage Match.* Students are provided with a set of cards that include different representation of rational number. These are used in a "Rummy" game where the different representations form a group. The cards can also be used for other matching games.

- *Build it up*. A group of four to six students have a collection of blocks. Each student is provided with a card that has information regarding the specific arrangement of a set of blocks they need to produce. Each has different information but all contributing to the solution. They are required to collaborate and use all the information to create a group solution.

Sourcing and Creating Purposeful Representational Tasks

As previously mentioned, a model or representation needs to connect directly to the mathematical concept that is the focus of the task. Such quality classroom tasks must move beyond a simple introduction to provide an engaging challenge to the students. In sourcing examples, we found ourselves returning to our favourite tasks that we had used over many years.

The creation of high-quality tasks of this form is challenging and in most cases, the tasks were based on the work of others. New opportunities did emerge with the use of ICT, particularly interactive whiteboards, for teachers to build tasks around visual representations. The most productive task adaptation or development came through teachers working in teams using the mathematical concepts or misconceptions as the starting point. A strong understanding of the mathematical concepts and related student understanding seem essential to producing quality tasks of this form.

Some Reactions from Project Teachers to Purposeful Representational Tasks

After developing, adapting, and using these tasks, the teachers completed a survey with a series of open prompts focusing on the specific task type. There were three groups of teachers with varying experience with other types of tasks in the context of the project when the survey was completed ($n=31$). However, there was little difference between these groups in the patterns of responses, except that teachers seemed to provide more extensive responses and include more comparison references if they had experienced the use of other task types. This is not surprising, as they were then in a position to note appropriate contrasts. The results from this survey are presented in the following sections.

The Teachers' Definitions of the Tasks

The first survey prompt was "if you were talking to a group of teachers about tasks of this type, how would you describe this type of task?"

The importance of the model and the explicit focus on the mathematics were the most common components of teachers' responses. The linking of the model or tool explicitly and directly to the mathematical concept was highlighted. Sample responses included the following:

> Using Models/tools representations to explicitly focus on a particular mathematical idea or concept. Often takes the form of teachers introducing a mathematics idea and students play a game or complete an activity. Follow up discussion on understanding/learning with students.
>
> It is a task where a "model" is used to explicitly teach the mathematical teaching. Models can be concrete materials, graphs, pictures, other representations.
>
> Using concrete materials, tools or representations to exemplify a mathematical concept, the tool can create a mental model for the kids.
>
> The use of a model or representation to assist student understanding of a particular mathematical concept to be used as a reference for further student work on the concept.

These teachers had been provided with some exemplars of tasks as well as having experience working in school- and cluster-based teams to develop and trial tasks. Their responses were consistent with the intentions of the researchers.

Some Examples of the Tasks that Teachers Valued

In the survey, teachers were asked "Of the tasks of this type that you have tried in your class this year, which worked best?" They were then asked to list the "next two best tasks".

Not only did no particular task emerge as the most popular, but also the most striking feature of the responses was the diversity of tasks that were valued, with 17 different tasks identified as working best across the 31 teachers. The Chocolate Block task, where the sharing of chocolate represents both an engaging context and a model for the development of the concept of fractions as division, was identified by the most number of teachers in their best three.

The reasons that the teachers gave for selecting the Chocolate Block task included

> Gave students something they could see. They were interested in the chocolate so it remains in most of their minds.
>
> ...was so effective in engaging the children and representing fractions as division.

The following quote, referring to the *Decimal Maze* task (see Chap. 13), was from a teacher responding to why it was successful:

> Concept that hasn't been introduced was made explicit through the use of this model. Students could see clearly what maths happens when you divide/multiply by a number larger/smaller than one.

The project teachers were teaching in the middle years with the vast majority in grades 5 and 6. An important curriculum focus in these years is fractions and decimals, and the majority of the best tasks of this kind focused on children's development in these areas. This would seem to be due in part not only to curriculum importance but also to the nature of the content and the availability of effective models and tools. Of the 81 nominations by the teachers as part of their top 3, 51 were focused directly on learning in number. This was despite the fact that in later professional development, an attempt was made to present and encourage teachers to try these kinds of tasks in other content areas. Teachers tended to trial and rate highly those tasks that they had experienced during project meetings or developed as a team within the school. As is discussed later in this chapter, teachers found it difficult to develop their own tasks of this type.

There was limited justification for the teachers' preferences for tasks, but the key themes appeared to be the engagement of the children followed by the importance of the model/tool/representation in enabling mathematics learning.

The Advantages of Purposeful Representational Tasks as Seen by the Teachers

To gain insights into the opportunities and advantages of the specific task type, the prompt was "What do you see as the advantages of using this task type in your teaching?"

The most common feature in the teachers' responses was the value of these tasks for developing student understanding. There were also many who commented on the engagement of students both in the sense of participation and in the way the model (sometimes referred to as "visual" or "hands-on") allowed engagement with the mathematics.

Increasingly during the first phase of the project where the different task types were trialled, the discussion of the teachers involved the role of different types of tasks and the value of using a range of types of tasks in their planning. The following quotes about the advantages of purposeful representational tasks were from teachers who had also trialled open-ended and context-based tasks:

> Yes as a starter to teach new maths that then can progress to [open-ended tasks].
>
> Having something concrete that the students can work with. Often in a game/interesting/engaging format for students. You can scaffold and work with students—questioning and guiding them
>
> In particular areas of maths eg—using operations—using this task allows students to learn the maths skills required before moving into applying it in a variety of contexts.
>
> It can support the concrete understanding of a maths concept for students for whom more abstract mathematical understanding may not develop as readily.

The Constraints on the Use of Purposeful Representational Tasks as Seen by the Teachers

The following prompt was included to provide insights into the constraints that teachers identified: "What makes teaching using this task type difficult? What are the challenges in using this type of task?"

The difficulties that appear to be related directly to teaching purposeful representational tasks include the difficulty in identification of these kinds of tasks within particular content area (e.g. chance and data) and the time required to prepare the materials.

> Sometimes finding the task. For me sometimes deciding which task is actually a [purposeful representational task]. Finding the resource and preparing it for use with a grade can be difficult ie time needed to copy, laminate, cut, etc.
> Ensuring each student has sufficient background knowledge and skills. At times I found it difficult to make the task relevant to the maths program.
> Understanding the model/representation is most effective when there is a purpose, ie our opportunity to apply their understanding of the model in a meaningful context. Also, particular maths concepts are easier to find models for.
> Sometimes modifying for lower students.

Some of the challenges of teaching such tasks are illustrated in the following series of classroom stories. They emerged from observations and interviews as part of the implementation of teaching sequences, one of which is described in Chap. 8.

The Challenges of Teaching Purposeful Representational Tasks: Learning to Use a Ratio Table

To illustrate some of the challenges in teaching with purposeful representational tasks, the following is a discussion of the use of ratio tables in a number of connected tasks.

The ratio table was developed as a tool for improving the understanding of ratio and proportion in Dutch primary and secondary mathematics schools (Broekman, Van der Valk, & Wijers, 2000). It is intended to assist students to develop mental strategies for solving proportion problems and particularly through the use of strategies such as halving, doubling, and multiplying by 10 (Dole, 2008; Middleton & van den Heuvvel-Panhuizen, 1995).

Broekman et al. 2000 identified the following six characteristics of a ratio table:

- The table consists of two (or more) rows and two or more columns, with numbers in the cells.
- The rows have a label, indicating the meaning of the numbers and specifying, if needed, the units used.
- The ratio between the numbers in the cells of a column is the same for all columns; this can be used to calculate an empty place in a column.

- There is no preference on what to choose as the upper or lower row.
- To get the numbers of a column, the numbers of another column can be multiplied or divided by a certain number. (Proportionately adding or subtracting is possible as well).
- There is no prescribed series of steps; pupils are free to follow a strategy on their own level (p. 17).

As part of a teaching sequence developed and implemented by teachers at one of the project primary schools, the use of ratio tables was introduced. The research team provided limited advice on the use of these tables and the following classroom snippets trace one teacher's (Ms A's) evolving understanding of the use of this tool through a series of tasks. The discussion also highlights some of the challenges of using such tools. The notes were from a lesson observation.

Making Cordial: Taking Opportunities

At the start of the lesson, Ms A introduced the making of cordial and the ratio of cordial to water through demonstration and discussion. It was an engaging and focused discussion.

The intent was to determine the ratio of water to syrup that created cordial that was "just right". Ms A then introduced the ratio table to represent the proportions as she increased the amount of drink produced. The table written on the whiteboard was a summary of the activity:

Cordial syrup	1	2	3
Water	3	6	9

Ms A's mathematical intent was to focus on the multiplicative relationship, but there was no attempt to extend the table or the associated questioning that might have facilitated the appropriate thinking.

It was not clear whether this was a missed opportunity or a specific instructional decision. The possibilities of the use of this ratio table for further learning were not taken in this case.

Bottles Task: Accepting Different Strategies

In a subsequent session, a group of children was given the following worksheet. It involved them in filling out ratio tables and answering questions exploring the relationship between the number of bottles of mineral water and the number of cases of bottles.

The group of nine students was sitting around a table with Ms A working with two of the students (Michael and Susan).

Name ______________________________ **Date** ______________

Section B. The Ratio Table

Bottles (Page 1 of 2)

1. Camp's Mineral Water is shipped in cases. Trudy Camp, owner of the company, uses the table below to keep track of the number of cases she has to ship each time a customer orders a certain number of bottles. Each case holds 15 bottles. Find out how many bottles there are in two, three, four, or more cases. Write your answers in the table below.

Cases	1	2	3	4	5	6	7	8
Bottles	15							

2. Jake's convenience store orders 90 bottles of mineral water. How many cases should be shipped?

3. Jake orders 65 bottles of mineral water for a work party. How many cases is this?

4. Camp's Seltzer comes in a larger case that holds more than 15 bottles. Trudy Camp's table for seltzer was ruined when she spilled water on it. The table below shows the few numbers that Trudy can still read. Can you tell from the table how many bottles there are in each case? Explain your answer.

New Cases					5	6						12
Bottles					125	150						300

5. How many bottles would fit in 18 of the larger seltzer cases?

32 Section B The Ratio Table Britannica Mathematics System

The last question on the worksheet was "How many bottles would fit in 18 of the larger seltzer cases?" Michael had extended the ratio table from the earlier question to find the answer (see below), while Susan had done a multiplication calculation using the standard algorithm (25×18) to answer this question.

New cases	1	2	3	4	5	6	7	8	9	10	11	12	13	14	15	16	17	18
Bottles	25	50	75	100	125	150	175	200	225	250	275	300	325	350	375	400	425	450

Ms A asked Michael how he had worked it out and then asked Susan to explain her method to him. Ms A commented in the post-lesson interview:

> Susan understood straight away, you could just go 18 times 25. But I was trying to get Michael to see it, but he.....

In justifying his answer to question 5, Michael had explained that for four crates there were 100 bottles so four lots of four, which is 16, is 400 bottles and then for the extra crates that would be 50, seeming to evidence multiplicative thinking beyond the apparent additive method from his table in number 4. Ms A seemed to be interpreting his method as solely additive.

> And I wanted him to see that one case is 25 bottles, two cases is 50 bottles, three—and see the picture of it so that he could see that it's three times, or two times, or one times. Which he did get to but he was adding them all.

After considerable discussion including the following notes made by Ms A, Michael was not convinced about the need to change his method. He said that they were both right and he preferred his. Ms A accepted this.

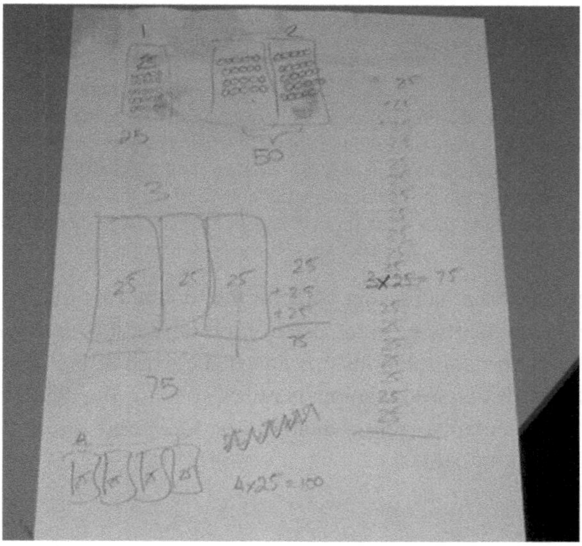

Ms A has clearly provided a range of ways in which 18×25 could be represented and interpreted but independently of the table of which Michael is trying to make sense. The use of ratio tables is intended to support proportional thinking and the tool lends itself to the thinking that Michael evidenced. If we are to use such tools effectively, then ideally a teacher needs to use them to link to the more abstract process he/she is intending to develop and not move too quickly to the "abstraction". While there is clear evidence in Ms A's notes above of her knowledge of a range of representations of multiplication, the link to the ratio table is not evidenced.

Ms A's approach may not have supported productively Michael's developing understanding of ratio. The "correct way" is presented as an explanation of 18×25

rather than the building to the number of bottles in 18 cases through alternate but acceptable methods linked to the ratio table. Knowing when to encourage the abstraction of the mathematics from the model can be challenging and is an important aspect of the pedagogical reasoning inherent in the use of a purposeful representational task.

Medals at the Shrine: Operationalising the Tool

Later in the teaching sequence, Ms A spontaneously used a ratio table to represent a ratio to the class. She made the following explanation in the post-observation interview:

> Because we'd been doing a lot of ratio table beforehand, and on Friday we went to the Shrine of Remembrance, and he said "there's 4000 medals here, every medal there's ten people who went and six who died", and I thought I'm going to use that as a little introductory, just to show them that you can use it everywhere, and a table is a really helpful way to work it out.

A later interview suggested that Ms A was increasingly seeing the value of the tool.

> And I've started to notice … they're using ratio tables to help them work out some of the questions. The one where Tom makes scones using a ratio of 1 cup of flour to 2 cups of milk, or something like that, and some of them have used a ratio table to work out for the class, and then they jumped to the levels. So we were talking last time you were here, we talked about them using the table, but not being able to make the jump; they got from 1, 2, 3 up to 10, but then [now] they work it out if it goes to 100 …And they are doing that now, which is good.

Ms A appeared to see increasing value in the tool and its true potential for supporting mathematical learning. In the use of ratio tables, it is necessary to highlight the horizontal and vertical relationships and make explicit the appropriate mathematics. This series of classroom snippets raises some of the challenges in the successful use of mathematical representations or tools and the implementation of purposeful representational tasks.

Summary

In the trialling phase of this project, there was a number of issues that arose in the teaching when using purposeful representational tasks. The teacher quotes and summary comments as well as classroom stories mentioned earlier provide some insights into those issues. They can be summarised as follows:

- While these tasks are not contextualised, there is sometimes a "hook" that helps to engage the students.
- Some content areas, particularly number, seem to provide more opportunity for successful purposeful representational tasks.

- Extensive exposition is not necessarily required. The provision of the model or representation can usually enable the students to generate the mathematical ideas and justification for themselves.
- The model, representation, or tool is best linked closely to the mathematical concept being developed.
- The mathematical focus is pivotal and it seems that teachers might be less willing to deviate from the intent than with contextualised tasks, for example.
- The tool or model needs to be seen as an important cognitive support and abstraction should not be rushed.

The teachers in this project were able to articulate the purpose, opportunities, and constraints of these tasks. Such tasks are an important component of curriculum, but developing them and teaching them well are not simple. However, as one of the teachers pointed out, *it is important to be reminded why* we need to include them.

Chapter 5
Using Mathematical Tasks Arising from Contexts

This chapter describes the second of the types of tasks that we focused on in our project. We argue consistently throughout this book that a teacher's choice and use of tasks are a major determinant of the nature and quality of student learning in the mathematics classroom. In this chapter, we present data from teachers indicating that the use of tasks built around practical (or "real") contexts can make mathematics "come alive" for students through showing them a purpose for what they are studying, and making mathematics more engaging for them. Within this chapter, we refer to them as contextualised tasks. Within the TTML project, it was assumed that the teacher would pose the task, clarify terms, context, and purpose, but would not tell the students what to do or how to do it. The teacher would orchestrate a class discussion after students had engaged with the task to hear interesting responses that teachers had specifically identified while the students were working, and would seek to draw out commonalities and generalisations. We also discuss the ways in which three teachers used a single task in quite different ways and how the influence of their individual mathematical confidence and preparation for a lesson yielded very different responses from the students.

A Rationale for Tasks Built Around Practical Contexts

In the 1980s, there was a strong movement around the world for school mathematics to focus more on applications. This was in part a reaction to the emphasis on pure mathematics in the new mathematics movement (de Lange, 1996). Niss (1987) argued that the increasing importance of mathematics in people's professions and everyday lives in an increasingly technological society was the compelling reason for its place in the curriculum. There was also a recognition that rather than starting from more abstract concepts and then leading to applications, contexts should be the starting points (Keitel, 1993). This view is not new. For example, Thorndike (1926) proposed that "the school should set problems in arithmetic which life then and later

P. Sullivan et al., *Teaching with Tasks for Effective Mathematics Learning*,
Mathematics Teacher Education 9, DOI 10.1007/978-1-4614-4681-1_5,

will set, should favor the situations which life itself offers and responses which life itself demands" (p. 12).

Those who support the use of contextual tasks suggest benefits such as enhanced motivation of students, as they see the ways in which mathematics can help us make sense of the world (see e.g. Meyer, Dekker, Querelle, & Reys, 2001). Proponents of *Realistic Mathematics Education* (RME) from the Netherlands advocate the use of contextualised tasks, but rather than emphasising motivation, for example, they focus on providing learning situations that are experientially real for students and a springboard for advancing understanding (Gravemeijer, 1997). In their view, the task can require students to "imagine the situation or event so that they can make use of their own experience and knowledge" (van den Heuvel-Panhuizen, 2005, p. 3).

The support for contextual tasks is not unqualified however. Some writers (e.g. Lubienski, 2000) have found that contexts can obscure the mathematical purpose for some students, particularly those from low SES backgrounds. For this reason, Sullivan, Zevenbergen, and Mousley (2002) called for more critical task selection and implementation. They and other authors (e.g. van den Heuvel-Panhuizen, 2005) encouraged teachers to avoid tasks which simplified practical situations unrealistically or used situations unfamiliar culturally to many students, or used mathematics to solve problems in unrealistic ways. Peter-Koop (2004) summarised many of the difficulties that students face when solving context-based problems, including comprehension of the text and the identification of the mathematical core of the problem.

Lovitt and Doug Clarke (1988) coordinated the Australian *Mathematics Curriculum and Teaching Program* in the 1980s. They asked teachers to identify their *concerns* about the teaching of mathematics. Common responses can be summarised as follows: mathematics was seen by many students as boring and irrelevant; little thinking was involved; the subject was too abstract; there was a fear of failure; teachers "covered" too much content in too little depth; assessment was narrow; and it was a huge challenge to meet the needs of a wide range of abilities.

Recent reports indicate that these concerns and others remain continuing issues in middle years' mathematics. For example, the Executive Summary of *Beyond the Middle* (Luke et al., 2003), a report commissioned by the Australian Department of Education, Science and Training, and involving a literature review, a curriculum/policy mapping exercise, and system, school, and classroom visits, included the following statement:

> There needs to be a more systematic emphasis on intellectual demand and student engagement in mainstream pedagogy. … This will require a much stronger emphasis on quality and diversity of pedagogy, on the spread of mainstreaming of approaches to teaching and learning that stress higher order thinking and critical literacy, greater depth of knowledge and understanding and increases in overall intellectual demand and expectations of middle years students. (p. 5)

Similarly, the Third International Mathematics and Science Study (TIMSS) Video Study, of particular relevance to this chapter, reported that just over a quarter of problems in Australian classrooms were set up with use of real-life connections, compared to 42% in the Netherlands, for example (Hollingsworth, Lokan, & McCrae, 2003). This would appear to indicate that many teachers are not regularly using contexts to build engagements.

Although the contexts used in this project are in some cases somewhat contrived, it is important to distinguish contextualised tasks from *word problems* (e.g. Fennema, Franke, Carpenter, & Carey, 1993), which are only contextualised in a basic way. A problem of the kind, "If oranges are $2.70 per kilogram, how much would four kilograms cost?", although placed in a shopping context, is not the kind of task we focused on in our project.

On the theme of students engaging with context, a frequently cited word problem (an army bus holds 36 soldiers. If 1,128 soldiers are being bussed to their training site, how many buses are needed?) brings up the issue of students apparently disregarding the reality of the situation (Carpenter, Lindquist, Matthews, & Silver, 1983). Of the 70% who correctly got a quotient of 31 remainder 12, only 23% gave the final answer as 32 buses, indicating that most students, far from becoming engaged by the context, completely disregarded it. Greer agreed that such word problems can actually teach students to suspend real-world sense making, but argued that "a case can be made for their retention, provided they are appropriately reformed, embedded in different styles of classroom interaction, and mixed with other forms of activity" (p. 299).

Greer (1997) noted that in a study by Curcio and DeFranko, a variant of the bus problem involved students making a telephone call to order the transportation, and this yielded 85% appropriate responses. This indicates that when tasks are made more "real" for students, they are quite capable of responding appropriately. He recommended that teachers treat word problems as exercises in modelling or mathematisation, viewing modelling as "the link between the 'two faces' of mathematics, namely its grounding in aspects of reality, and the development of abstract formal structures" (p. 300). Greer (1997) called for the interpretation of a given word problem to be as a "lean description of some situation, the details of which must be filled out by plausible (but debatable) assumptions, sensitive to many aspects, such as the goals of the problem solver" (p. 297).

In discussing contextualised tasks in this book, we use examples which are intended to be related to the interests and experiences of middle-school students, while avoiding what Maier (1991) described as school problems coated with a thin veneer of "real-world" associations.

Hodge, Visnovska, Zhao, and Cobb (2007) studied the use of a range of contextualised tasks with seventh-grade students in the USA, with a focus on the extent to which these tasks supported students' mathematical engagement and their developing mathematical competence. Most of their tasks involved comparing two data sets in order to make a decision or judgement (e.g. deciding whether the installation of airbags in cars impacts on car safety, exploring the impact of a treatment programme for AIDS patients). During the design experiments, the authors found that issues which were of a personal or societal relevance were the most effective in engaging students. They attributed this to "adolescents' growing interest in their place in society and their sense of power in affecting [sic] change on society and their immediate community" (p. 398).

There is, of course, a need to take care in deciding what constitutes relevance for students. Middleton and Jansen (2011) offered a cautionary note about a US teacher who used a tipping context to justify calculating 15% of a given amount, only to discover that few of her students (or their parents) would ever be in the situation where such calculations were necessary, given their family income levels. They argued for "judicious use of contexts" (p. 110), and offered several strategies to address the challenges of making contexts accessible and engaging to students:

> Teachers can provide scaffolding to help students learn about the story problem's context as well as its mathematical relationships. Students could generate their own contexts. Teachers could give students choices about which contexts to use for a task. Teachers could learn about students' communities and neighborhoods to design problem contexts that include locally valuable and relevant situations. (Middleton & Jansen, 2011, p. 110)

So, there are many challenges facing teachers, schools, and systems in improving both cognitive and affective aspects of students' mathematics learning in the middle years. We argue that contextualised tasks have great potential for challenging and engaging students, and showing how mathematics can help them to make sense of the world.

The Contextualised Tasks That Were the Focus of Our Project

When using contextualised tasks as part of this project, teachers situate mathematics within a practical problem where the motive is explicitly mathematics. These tasks have a particular mathematical focus as the starting point and the context exemplifies this. The context serves the twin purposes of showing how mathematics is used to make sense of the world and motivating students to solve the task. An example of this is "How many people can stand in your classroom?" (Lovitt & Clarke, 1988) where the task posed is of the kind, "Imagine we have the opportunity to put on a concert in this classroom with a local band to raise funds for more school computers. How many tickets should we sell?"

Here, the context provides motivation and engagement for what follows and dictates the mathematical decisions that the students make in finding a solution. The teacher will have broad intentions, in advance, about how the content relates to aspects of mathematics, specifically, an understanding of area, estimation strategies, and the notion of measurement errors.

In seeking to use contextualised tasks then, a teacher needs to choose tasks that connect with the interests and experiences of students, address important mathematics, engage students, and help them to see mathematics as making sense of the world. We now discuss in some detail the use of a task which arguably met all these criteria.

A Specific Example of a Contextualised Task

A number of teachers in our project used what we came to call the *Signpost Task*. The photos within this chapter and the description are based largely on the use of the task by one project teacher (see Doug Clarke & Roche, 2009). The tasks involved students conjecturing where the photo shown in Fig. 5.1 was taken.

It was predicted that this would involve students in the use of scale, estimation, and an informal introduction to the idea of locus, in relation to all the possible cities which could be, for example, 2,159 km from Sydney.

Setting the Scene

The teacher asked students whether, during family travels, they had ever seen a sign at lookouts or at other tourist places which showed how far and in which direction a number of key places were from their current location. Most had seen such a sign (mostly at lookouts), and they shared their experiences.

The teacher then held up a picture (as shown) which had been taken of such a sign, showing the distances in kilometres of 14 other cities from the signpost, and explained that that day's lesson would involve the students working, in pairs, on trying to find out the location of the signpost. Students were asked to offer initial thoughts on the location prior to setting them to work. The somewhat "tropical" background can often be something of a distraction in their predictions.

The teacher then indicated that students were free to work on this problem in pairs, in any way they wished. The students chose their own partners. Atlases were provided for each pair, and rulers and calculators were available if required.

Enabling Prompts

While most pairs decided upon a starting strategy and got to work, several students seemed unable to make a start on solving the problem and required some assistance. Sullivan, Mousley, and Zevenbergen (2004) coined the term *enabling prompts*

Fig. 5.1 Where was this photo taken?

(as discussed in Chap. 3) to refer to appropriate variations on the task or suggestions to students which might help those who are having trouble making a start on the problem. One helpful enabling prompt in this case was to suggest to students that they pick a city named on the sign and find out how far on the map it would be from the sign's location and therefore which "mystery city" might contain this signpost.

This prompt seemed helpful, but it still provided a challenge, as it involved students using a scale to see how the distances on the map related to the real distances in kilometres. For example, if Seoul is 9,636 km away from the signpost, what would this be in centimetres? Some students used the scale below by measuring with their ruler the length from zero to 4,000, doubling this length to make 8,000 km, then adding the length from zero to round about 1,600 to make a length that closely represented the distance to Seoul.

Having come up with an approximation, they then measured in a straight line from Seoul that many centimetres in a variety of directions to establish possible locations for the signpost.

Students' Strategies

Students varied in their abilities to read and use a scale, and to convert the distances shown on the signpost to centimetres on the map. In discussion with teachers, it was clear that students in the middle years rarely are asked to either create or use a scale, and that scale probably needs greater attention in the intended and enacted curricula.

For some students, the enabling prompt was not necessary as they had noted that Sydney and Fiji were almost equidistant from the signpost (2,159 km and 2,157 km respectively) and the closest places therefore to the "mystery city".

On one world map, the scale was expressed as "1 cm on the map represents 450 km on the ground". Some students needed assistance to see that the calculation of $2{,}159 \div 450$ would give an approximate distance on the map from Sydney to the mystery city. Explaining that this equation is solving "how many lots of 450 would go into 2,159?" seemed to be helpful. Students then were encouraged to estimate this ("about 4 or 5") and then to find the answer on a calculator.

Some difficulties arose around the need to be accurate when converting and measuring, particularly when working with some of the larger distances. Students were free to choose any cities and atlas maps to work on, but not all of the signpost cities (or the mystery city) were on the maps which students chose. This meant that students had to estimate or even guess where their cities might be. Students sometimes needed to move between maps of the world and maps of the Pacific region, once they realised that the mystery city was likely to be somewhere not far from Australia.

Building upon Students' Insights

One pair of students had worked out that the mystery city was close to 5 cm from Fiji on the map. The pair then showed the teacher their insight by rotating their ruler around Fiji, and noting where 5 cm from Fiji "reached". They knew that the city must be somewhere along this imaginary circle.

The teacher took this opportunity to share the insight with the class and introduced the use of a mathematical compass for construction, in order to make the measurements more accurate. It was not long before some students realised that drawing a very similar-sized circle around Sydney would provide other important information.

The challenge for students then was to take these two circles and decide what it told them. The first student to speak decided that it must be within the region created by the overlap of the two circles, and this seemed to draw agreement from some other students. The teacher pointed to a spot within the region, and asked the students whether this point was 5 cm from Fiji. After some further discussion, there was a consensus that the mystery city must be close to one of the two intersection points of the two circles.

The students then excitedly used other information about distances to establish that it must be the southernmost of the two points, leading to an answer of somewhere in New Zealand.

Pulling the Lesson Together

In most of the classes where this task was used, the teacher called upon a small number of pairs to share their reasoning. The pairs were generally chosen to represent a variety of different approaches and/or challenges faced. In several classes, teachers asked the students to talk about the mathematics they had learned (e.g. creating and using scales, careful measurement, using a compass to create circles, estimating, predicting, and checking). Some students also noted that strictly speaking, just two distances were enough to narrow down the possible cities to two (the intersection points).

Students as Problem Posers

A number of key writers in the problem-solving area (e.g. Brown & Walter, 1993; English, 1998; Silver, 1994) have stressed the importance of *problem posing* by students. Several teachers in the project took the opportunity in subsequent lessons to extend the work on the task, by encouraging students in groups to create their own signposts with cities of their own choice, and then to pose their problems to another group. This was an excellent way of consolidating the learning that had taken place during their work on the original task. Students appreciated being able to choose their own cities, and undertook the task enthusiastically.

Some Examples of Contextualised Tasks Which Teachers Valued

In the survey, teachers were asked, "of the tasks of this type that you have tried in your class this year, which 'worked best'". They were also invited to describe the "next two best tasks".

The nine tasks on the following pages include all tasks mentioned by at least one teacher. The nine tasks provide a mixture of contexts in which mathematics can be found. In the case of Soccer's New Dimension and Mike and His Numbers, the context is drawn from a newspaper article. In the case of Seat Belt Sampling and Petrol Card Choices, a photograph taken in the outdoors has provided the context. With What Time Does This Clock Show? and Maps for the Commander, the context is somewhat artificial, but nevertheless engaging to most students. The Music Cards task draws upon students' interest and experience in downloading songs from the Internet. The Block of Land comes from a mathematical request to one of the authors, and the D-Tape provides a context in which an understanding of circumferences of circles is used in the forestry industry.

Music Cards

Which card is better value?

Pod Tunes
16 songs $24

New Tunes
12 songs $20

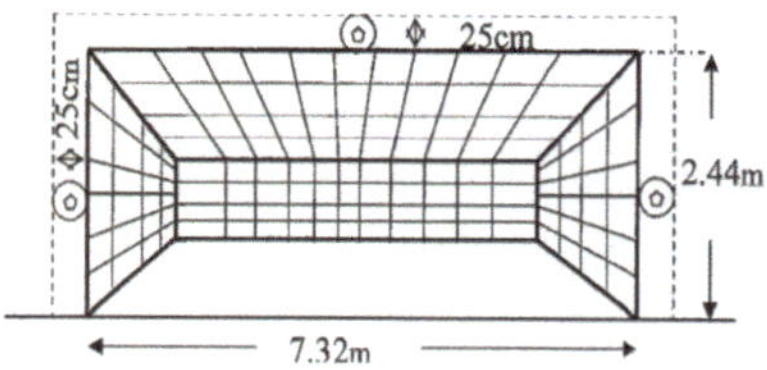

The Size of Soccer Goals

At one time, it was proposed to increase the number of goals scored in soccer by extending the soccer goal by the width of a soccer ball, on three sides, as shown. Investigate the likely increase in goals which would result

---

Maps for the Commander

An army commander sent three spies to draw different views of a city which was surrounded by a circular wall. Two spies returned (see their sketches at right), but the not the spy who sketched a view from the north-east. Please draw that sketch (adapted from work of Jan de Lange)

View from the west View from the south

1, 2, 3, 4, 5, 6, 7, 8, 9, 10, 11 ….1,000,000

Mike and His Numbers

Mike from Devonport in Tasmania wrote down every number from 1 to 1,000,000 in his spare time over a two-year period in the 1980s. At the time, one local newspaper reported that the total number of digits Mike wrote was 5,878,936

Students are asked to investigate whether the newspaper's claimed number of digits was correct or not

---

(continued)

(continued)

Seat Belt Sampling

Assuming that the same number of cars is sampled each month (and the result rounded to the nearest percent), what is the smallest sample size which could produce these figures?

Diameter Tape

A diameter tape (D tape) is used by foresters to measure the diameter of a tree. The D tape is wrapped around a tree at around breast height and converts the tree circumference to tree diameter

Using a streamer, ruler, and calculator, make your own D-tape. You will be asked to use it to read off the diameter of a "tree" with a diameter between 40 and 50 cm

---

What Time Does this Clock Show?

This clock has been dropped on the floor and we don't know which way round it should be. What time is the clock showing? (adapted from the idea of Peter Gallin)

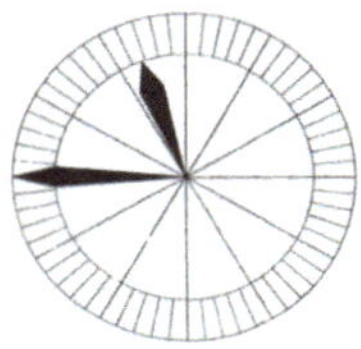

5% off BP Petrol
(beats 4c off per litre)

Petrol Card Choices

This is a sign from a petrol station

When is 5% off better than 4 cents off per litre?

Block of Land

Suppose you received this letter from a friend:

What would your answer be?

Hi,

Can I call on your maths expertise?

If, on paper, a block of land is 2 cm x 5.8 cm, and the overall dimensions are 4768 m square, how do I work out the actual length and width of the block?

Hope you can help.

Some Reactions from Teachers to Contextualised Tasks

The project explored the nature of the learning based on contextualised tasks, the opportunities that such tasks offer to teachers and students, and the constraints that the tasks created for teachers in their use. After the teachers had worked with the respective task types, they completed a survey which asked them questions about these issues. Their responses were categorised and summarised. This enabled us to draw upon teachers' experiences and to see how what had been discussed in the literature played out in practice.

Teachers were asked to describe contextualised tasks as they would if explaining them to another teacher. The prompt was "If you were explaining to a group of teachers about to use tasks of this type, how would you describe this type of task?" Our aim in posing this question was to see whether the understanding by the project team of what distinguished these kinds of tasks was shared by the teachers.

The following examples are typical of the kinds of responses teachers made:

A mathematical problem embedded in a real situation.
Questions which allow/require investigation through use of materials, data gathering, testing and calculation.
The tasks are based in authenticity.
The mathematical problem is contextualised, but with an explicit maths focus.
Contextualised maths investigations with explicit mathematical focus.
Application tasks involving situating mathematics within a contextualised practical problem where the focus is explicitly mathematics.

The project team was pleased with the teachers' descriptions of contextualised tasks, as their responses indicated that most had a clear picture of the nature and purpose of these kinds of tasks.

Teachers' Views on Advantages and Difficulties in Using Contextualised Tasks

After at least one school term of trialling a range of contextualised tasks, teachers were asked to list "advantages of using this task type in your teaching". Typical comments were as follows:

More hands on.
Some were good for the students who struggle with mathematics.
The mathematical skills and strategies are made purposeful and meaningful by being situated in a "real world" context.
Increases the students' ability to think.
Allows the students to draw on a variety of understandings and topics—engaging and relevant to what they are doing.
Engages advanced students. Combines knowledge and skills, e.g. a task may need measurement, calculation, logic.
Each task can be taken in various directions by the students. There are different ways to solve the puzzle and are very engaging.

In attempting to get at what constraints were experienced by teachers when using contextualised tasks, teachers were also asked "what makes teaching this task type difficult?" In the comments below, "support students" refers to those students in the classes in which students of "lower ability" were grouped. Representative responses were the following:

> Some of the tasks were too challenging for support students and too long!
> The different learning needs and abilities of the students; at times some students arrived at their conclusions more quickly then others.
> Students who are less confident have very little idea of where to start if left to their own devices rather than assisted. These tasks can compound their negative feelings about themselves and maths.
> Not all the real situations are relevant to middle years students and may not fit neatly into the existing curriculum.
> You need to do some preparation with the students. Students are more interested in the answer than the process.

It is worth noting that teachers in secondary schools found the contextualised tasks more problematic to use generally than did those in primary schools. Reasons given often related to the challenge for students described as "weak-" or "low-ability students" in making a start on such problems. Teachers also noted that the experiences of many students in junior secondary mathematics involved little exposure to tasks of this kind, and the required thinking and planning on their part in working on these tasks contrasted greatly with the repetitious exercises with which they were more familiar.

Teacher Actions and Their Impact on Task Potential and Student Learning

In this section, we discuss the ways in which a single task was used (or adapted) by three teachers. We see the way in which each teacher puts their own "spin" on the task, influenced by their concerns about likely student difficulties, and possibly their own mathematical knowledge and confidence. We see particularly how important it is that the teachers have worked through such a task themselves in advance of the lesson so that they can maximise the mathematical and pedagogical possibilities within it.

Three teachers in the one school taught a lesson based on the same task:

> Usain Bolt ran the 100 m. in 9.7 sec. How fast is that in km/hr? How fast you can run in km/hr?

The following are summaries of the lessons as derived from the teacher interviews and observations, with some interpretative comments.

Teacher A

Ms A's written lesson plan showed she intended to have a brief, initial discussion linking to previous lessons, a whole class discussion on km/h, after which the students would work outside in pairs on the task, followed by a whole class debrief.

As part of the 26-min introductory discussion, Ms A, an early career teacher with confidence in her ability and mathematical knowledge, posed the following problem:

> Stompy (the class turtle) escaped. He covered 10 metres in 30 seconds. How fast is this in km/hr?

Note that the *Turtle* question is of a different form from the *Usain Bolt* question, with the number of seconds being a relatively "nice" fraction of an hour. After working on this problem, one student wondered whether he could walk that fast. Ms A adjusted her plan to facilitate this incidental opportunity.

The students were then asked to work out how fast they could run. Before the students went outside, they were given detailed directions on organisational matters (e.g. using the stop watches), but no instruction on how to do the running task and related calculation.

The students spent 30 min outside. The students worked in groups, with some choosing to measure *how far they could run in a set time*. When asked, the students in those groups said that they chose their method deliberately since it would be easier to calculate. For example, one student said:

> [student name] and me chose to do ten seconds because if you do ten seconds, it needs to add to one hour but the distance doesn't really have to add to anything…

Other groups measured how long it took them to run a particular distance.

When the students returned to class, they continued to work in groups. Most of the eight or so students who had chosen the easier method calculated their answer readily. Many other students who had chosen the more difficult method struggled with the calculation. The teacher was extremely busy trying to help the students working on the difficult method, while the more capable students completed the work quickly (but pretended they had not yet finished). This phase took 25 min. There was no concluding review, and therefore no discussion of the differing methods.

In the post-lesson interview, Ms A recognised what had happened:

> So those that had thought about time and a unit of time prior to it were able to do it more readily than those that had thought about a unit of distance. So if I was to do it again … I would try to make the specific ratio idea clearer… I always try to put it back on them.

Ms A had thought about the task and its pedagogical purpose, and gave the students ample opportunity to devise their solution path for themselves. The task was well introduced, with the Turtle question being meaningful to the students, and at a lower level of difficulty. Ms A had not anticipated the way that the form of the calculation chosen determines the level of difficulty, although she realised this during the lesson. The task, and this lesson, clearly created opportunities for students and most students were able to respond to the assessment task. Interestingly, in a questionnaire in which students were asked to rate the various tasks in the unit, nearly half of the class chose this as the task they either most liked or from which they felt they learned most during the unit.

Ms A neither moderated the level of challenge for the students nor reduced the potential of the task, and certainly maintained a commitment to students choosing their own methods.

Teacher B

Ms B, a confident early career teacher, who was uncertain about aspects of her mathematics knowledge, described her plan prior to the lesson:

> …then we're going outside to calculate their kilometres per hour and that would be quite hard for some of them. So I'll just have to see how it goes with how far we get. The timing will be easy because we'll just time 100 metres and then convert it.

Ms B posed the task as "how fast can you run a kilometre?" She invited the students to suggest how they might do this. Various considerations such as the ability to maintain running speed were proposed. One student suggested "we could do 100 metres and times by 10" and this idea was adopted. Interestingly, other methods proposed by students were rejected quickly, as they appeared not to conform to Ms B's plan. One student asked whether they could do three sprints and find the average, and this was confirmed as a good idea by Ms B:

"How you work it out is up to you, but you'll need to share the trundle wheel and stopwatches. Work out as a group how you're going to record results. When you've worked that out you can come and get a stopwatch and a trundle wheel off me. So groups of three or four would be best".

The students then spent 25 min outside on the sports field, working in small groups on their data collection. The last five minutes of this part of the lesson was back in the classroom, with the students together. There was a discussion on how they could find an average, with the teacher giving the instruction, "go back through your maths book and see if you can find how to do it". This part of the lesson concluded with the instruction:

Now think about how to change your time to how many kilometre per hour you can run. Take it home and talk to your parents about it. See if you can work out how to do it.

In the post-lesson interview, Ms B was asked about the outcome of the lesson:

> I think they'll all need a little bit of help but I think some of them will be able to work it out. A couple of them have already done the Bolt question where you convert his speed of how it takes him 9.7 seconds to run 100 metres. I've already worked with a couple of students to help them convert that into kilometres per hour. So they'll be able to use that information to help them.

In the following lesson, after an unrelated introduction, Ms B did a six-minute review of the "how fast can you run?" lesson. After discussing strategies for calculating their average speed in km/h, the students then worked individually or in pairs for five minutes on possible methods for calculating the speed. There was limited success. Ms B led a discussion about Usain Bolt's speed with questions like "how many times do you have to multiply the 9.7 to get to an hour?" She wrote on the board "$9.7\times?=60$", leading to a procedural presentation of a solution. She then repeated this with some of the students' times (e.g. 21 s), modelling the procedure and then asking them to work on their own answer. They worked on this for 25 min.

We assume that Ms B posed the introductory task, "how long would it take you to run a kilometre?" this way to make the solution of the main task easier, but it did not do this. The orientation of the teacher towards allowing students to make their own choices was evident in her posing the home-based continuation to the first part of the lesson. In contrast, she imposed a particular data collection process before the students started to take their measurements. In class, Ms B's attempt to simplify the task and the direct and formal way that she presented a solution method were both planned, and perhaps limited, by her lack of confidence with the mathematics underlying the task. In other words, she attempted to reduce the potential challenge for the students, having anticipated student difficulties, and this seems connected to her own lack of confidence with the task itself. As it happens, her attempt to make the problem simpler for the students actually made it more procedural and more complex.

Teacher C

Ms C was a very experienced teacher, but one who expressed greater confidence in the teaching of literacy than mathematics. The lesson of Ms C, in short, was similar to that of Ms B, but different in three major ways. She showed a video of the actual race, she spent time in the introduction and conclusion on calculating time differences of different runners in the race (which was irrelevant to the ratio aspect of the lesson), and she drew reasonably skilfully on students' suggestions.

The challenges which the mathematics posed for Ms C were evident in the two questions which were posed to the students: "If he could keep up that same rate, how long would he take to run a kilometre? … How many kilometres per hour would he be running at if he could sustain that rate?" These two questions are getting at two quite different aspects: seconds/minutes per kilometre and kilometres per hour; and yet they were posed ten seconds apart, as though they were seeking the same outcome.

The observer noted later in the lesson:

> (Ms C) invited a student to explain his method to the class, and he said: "I got 100 metres and divided by 9.71, which gets me how much metres you got in a second, and then I multiplied by 3600." Ms C asked where the 3600 came from, and the student replied "60 seconds by 60 minutes," and gave the answer as 37075.18 km/hr. Ms C then restated the method suggested by the student, followed by a discussion of the need to divide by 1000 to change metres per hour into kilometres per hour.

In summary, Ms C had the intention of being explicit about the method the students should use, but drew the method from a student, before restating this method. After the lesson, she was aware that the method she chose was complex, and expressed a view that she would describe a simpler method another time. Ms C intended to restrict the student choice of method, and so reduced the potential of the task, and this seemed directly connected to her own lack of confidence in the mathematics needed to solve the task.

Some Reflections on the Three Lessons

The first part of the task that was the basis of these lessons is complex because of its real-world nature. Had Bolt run 100 m in 10 s, it would have been easier. The other part of the task (the running) was complex because of its openness and the student choice involved. Ms A did not moderate the demands of the task, although she did not use the *Usain Bolt* part. Both Ms B and Ms C intended to present a particular method of solution, apparently motivated by their lack of confidence with how to do the task themselves. At this level, knowing a formal method for rate conversions is of limited value, and the real potential of the task is the opportunity for students to work out a method for themselves. The reduction in the potential of the task by Ms B and Ms C was mainly in the restriction of the students' choices of the methods of solution.

The main message we took from these stories relates to teacher decision making being influenced by their knowledge and confidence with the mathematics, and is well summarised by Brophy (1991):

> Where (teacher) knowledge is more explicit, better connected, and more integrated, they will tend to teach the subject more dynamically, represent it in more varied ways, and encourage and respond fully to student comments and questions. Where their knowledge is limited, they will tend to depend on the text for content, de-emphasize interactive discourse in favor of seatwork assignments, and in general, portray the subject as a collection of static, factual knowledge. (p. 5)

The first part of the quote corresponds well to the actions of Ms A, while the second part hints at the perceived need of Ms B and Ms C to constrain student activity to a mathematical path with which they felt more confident.

In drawing out implications from the observations of these three lessons, we see a need for teachers to work through contextualised tasks they are planning to use in class individually at first, taking on the role of the learner. Ideally, they then discuss the mathematics with colleagues who have also worked through the same task, including discussing the kinds of methods and solutions which students might use and find respectively. They are then well placed to consider pedagogical implications in terms of how they will introduce the task, the ways of working which are expected, any adjustments to the tasks for students who struggle or those who find it straightforward, and how they will "pull the lesson" together at the end.

Sourcing and Creating Contextualised Tasks

There is a variety of ways in which contextualised tasks can be created or sourced. In this section, we provide a number of strategies for doing so, using specific examples of contextualised tasks in each case.

Starting with the interests of students. Many students feel that mathematics is irrelevant to their needs and interests, and classroom experience for many supports this view. In creating contextualised tasks, one strategy is to think of particular interests of students and attempt to develop tasks built around those contexts (Middleton & Jansen, 2011). For example, the task *Music Cards* was developed by our colleague Anne Roche, because she noted the widespread interest of students in downloading songs from the Internet. We had been developing a range of tasks around fractions and proportional reasoning, and realised that almost all of our tasks were devoid of context. In the same way, a teacher might develop a task based around various mobile phone plans, if they are of interest to students at a given time. Also, students are often interested in the life experiences of teachers, and so contexts that have arisen in teachers' lives may also provide sources of tasks. For example, the task *Block of Land* arose from an email sent to one of the authors.

Taking a particular event from the newspaper, television, or radio. Given that articles in newspapers almost invariably are based on contexts, they provide a ready source for contextualised tasks. For example, on October 9, 2010, *The Week* magazine reported that Britain is now more covered with trees than at any time since 1750, with 11.8% of the total land area being wooded. Students could take these data, estimate the percentage of their own country (e.g. Australia) that is covered, and then discuss how they could collect necessary data to compare with their estimates. There are many other examples in this chapter of newspaper or magazine articles providing the source for worthwhile tasks.

Taking digital photos of signs, or scenes of interest, with mathematical potential. We have learned that it is very worthwhile to have a digital camera with us (or very handy) most of the time. We take photographs of street signs (e.g. *Seat Belt Sampling*) which have mathematical potential, or advertising signs which talk about particular discounts. We have noticed that the more we get in the habit of taking such photos, the more opportunities we recognise.

Taking Prompts and Using Them to Develop Contextualised Tasks

In teacher workshops, we quite often provide samples of "prompts" for contextualised tasks, and invite teachers to talk about how they would use them in mathematics classrooms. For example, a prompt might be a table taken from a newspaper showing the percentage of people who claim to cry at certain events (e.g. sad movies and funerals). We are sometimes disappointed that teachers will say things like—"there is a table of data there—the students could graph it".

We usually respond with "but what is the question the students would be answering by graphing these data?" We believe that to make contextualised tasks clearly relevant to students' interests and mathematically worthwhile, we need a question to answer or problem to solve for focus.

If there is a particular mathematical focus and a clear problem to solve or question to investigate, it will help students to see that mathematics can help us make sense of the world, and they are less likely to feel that they are being given "busy work".

Summary

In later chapters, we provide more information on the use of contextualised tasks, teachers' and students' work with these tasks, and student preferences in relation to different kinds of tasks. While acknowledging that many teachers believe that the use of contextual tasks is challenging for them and some of their students, the claim here is that contextualised tasks can assist students to make connections between mathematics and its applications, and to see how mathematics can help to make sense of the world, in settings that engage most students. Teachers need to take care that the meaning and realism are not too contrived, and that allowances are made for students who might not be familiar with the context, considering enabling and extending prompts in advance. Teachers also need to take care that they are not using the context simply as a "hook", which is left behind when the mathematics becomes the focus. In pulling the lesson together, it is important that the teacher focuses on both the mathematics and what has been learned about the context.

When surveyed, teachers were generally positive about the use of contextualised tasks with their students, although there was no agreement among teachers that these tasks always met the needs of low-attaining students, particularly on the part of teachers of secondary students.

And Where Was the Photograph Taken?

The photograph was taken inside Auckland International Airport in New Zealand.

Chapter 6
Using Content-Specific Open-Ended Tasks

This chapter explains the ways that open-ended tasks might contribute to learning, it gives the details of a specific open-ended task and how it might be used in a "lesson", it indicates the challenges that teachers may experience when using such tasks, it presents a range of examples of this type of task to illustrate the scope and nature of the tasks, and it summarises some research on teachers' reactions to the tasks. The fundamental argument is that such tasks are accessible by students, able to be used readily by teachers, foster a range of mathematical actions, and contribute to some of the important goals of learning mathematics.

The Potential Contribution from Open-Ended Tasks to Student Learning

As discussed in Chap. 3, an assumption underpinning this book is that the nature of teaching and what students learn are defined largely by the tasks that form the basis of their actions. In this chapter, we discuss ways that working on open-ended tasks can support mathematics learning by fostering actions such as investigating, creating, problematising, communicating, generalising, and coming to understand procedures.

Also in Chap. 3, we explained our meaning for the terms task and activity. One further key term is the goal which is the result students seek as a product of their activity in response to the task statement. Goals can be open or closed. A closed goal implies there is only one acceptable response. A task has open goals when it has more than one (preferably many more than one) possible response and we call such tasks open-ended.

P. Sullivan et al., *Teaching with Tasks for Effective Mathematics Learning*, Mathematics Teacher Education 9, DOI 10.1007/978-1-4614-4681-1_6,

When working on open-ended tasks, students engage in actions different from those when they are working on closed tasks. Because open-ended tasks are generally less well defined than those that are closed, students are less prompted to recall a rule or procedures as a way of solving the task, and so need to consider the meaning of the concepts involved, make decisions about processes for undertaking the tasks, consider the possibility of multiple responses, and think about appropriate ways of communicating results. Students choosing their own emphases and approaches and developing their awareness of those choices are key elements of mathematics learning (Watson & Sullivan, 2008). The need for decision making by students is engaging since it increases their sense of control (Middleton, 1995).

Specific studies that support the use of open-ended tasks include that by Stein and Lane (1996) who noted that student performance gains were greater when "tasks were both set up and implemented to encourage use of multiple solution strategies, multiple representation and explanations" (p. 50). Likewise, Boaler (2002) compared programmes in two schools. In one school, the teachers based their teaching on open-ended tasks and in the other, traditional text-based approaches were used. After working on an "open, project based mathematics curriculum" (p. 246) in mixed ability groups, the relationship between social class and achievement was much weaker after 3 years, whereas the correlation between social class and achievement was still high in the school where teachers used traditional approaches. Further, the students in the school adopting open-ended approaches "attained significantly higher grades on a range of assessments, including the national examination" (p. 246). There have been those who have questioned these results, but Boaler (2008) has continued this line of research in different studies and has replicated the results in various places (see also Staples, 2008, for further confirmation of the success and processes of this approach).

There are various approaches to making classroom tasks open-ended. Wiliam (1998) described investigations such as "describe the bounciness of these balls". Leung (1998) outlined a related approach that emphasises problem posing. Christiansen and Walther (1986) argued that opening up tasks can engage students in productive exploration. They gave an example of "How much does it cost to keep a dog for a year?" While such tasks are important, our experience is that teachers experience difficulty in connecting such tasks to their classroom programmes which are constrained by mandated curricula and systemic external assessments.

In addition to the open-endedness described above, the open-ended tasks that we prefer are also content specific in that they address the type of mathematical topics that form the basis of textbooks and the conventional mathematics curriculum. Teachers can include such tasks as part of their teaching without jeopardising student performance on subsequent internal or external mathematics assessments. The definition that we used with our project teachers was as follows:

> Content specific open-ended tasks have multiple possible answers. They prompt insights into specific mathematics concepts through students seeing and discussing the range of possible answers and identifying the patterns in those answers. There is ideally choice in strategy and solution type, in that students might approach the tasks arithmetically, or they might seek more generalised solutions.

We also provided the teachers with specific advice on the pedagogies associated with this type of task. In particular we suggested the following:

> It is assumed that the teacher will pose the task, will clarify terms, will explain that there are multiple possible responses, but will not tell the students what to do or how to do it. The teacher will orchestrate a class discussion after students have engaged with the task to hear interesting responses (that teachers have specifically identified while the students are working), and will seek to draw out commonalities, and generalisations.

The following section illustrates how the dual aspects of such tasks, that they are open-ended and content specific, might be used to enhance student learning.

Creating a Learning Experience Around a Content-Specific Open-Ended Task

The following task is used to illustrate some key considerations for teachers. We would pose such a task to lower to middle secondary level classes. It is a simpler version of the common upper secondary problem of maximising the volume of a box created from a rectangular card. A closed version of the task would be:

> Squares of side 2 cm are cut from a rectangular sheet that is 20 cm long and 16 cm wide, and the resulting shape is folded into an open-top box. What is the volume of the box?

The open-ended version of the task is:

> Suppose that you have a rectangular sheet that is 20 cm long and 16 cm wide, and you cut squares out of each of the corners. You then fold up the sides to make an open-top box. Calculate the volume of some boxes that can be made from that card.

One advantage of the open-ended task over the closed version is that it is able to be used readily with heterogeneous classes because students can approach the task at different levels and in different ways. Open-ended tasks have been shown to be generally more accessible for a range of students than closed examples (see Sullivan, 1999).

Further, this form of the task can focus students' attention onto key aspects of concepts (when different sized squares are cut out from the corners, this results in different sized boxes, that the volume of these boxes varies, etc.) as distinct from them merely trying to recall and apply rules. These goals can be specifically stated for the students. The open-ended task also allows students opportunities to investigate the problem context, make decisions, generalise, seek connections, and identify alternatives. It also gives insight into the constraint on the size of the cut square leading to the notion of domain in later years.

Some readers may not see the latter task as open-ended since, by assigning a variable to the length of the side of the square, the volume of such boxes can be described "uniquely". We assume this task will be posed to students at an age prior to them having an orientation to seeking algebraic solutions to such tasks in this way. We anticipate that many lower to middle secondary students will seek numerical solutions, even restricting themselves to natural numbers. Nevertheless we anticipate that some students may explore the possibility of describing a general solution, especially after they have identified a range of specific solutions.

The first action is for the teacher to pose the task and make it explicit to the class what it is intended that the students learn. It is also important to recognise that some students may be unfamiliar with some linguistic or procedural aspects of such tasks. For example, the teacher could explore students' interpretations of key terms, such as *volume* and *open-top* box. The teacher could emphasise aspects of ways of working on the tasks such as that there are multiple possible solutions and many different ways of representing solutions; that creativity is desirable; that responses do not need to be presented neatly in the first instance; that it can help if they imagine what the box might look like; and so on.

The key at this stage is for the teacher to avoid telling students how to do the task and to let them explore the task for themselves. In classroom observations, many project teachers seemed tempted to reduce the task demand by instructing students on how to complete the task. It is ideal for teacher to resist the desire to do this.

The next stage is for the students to commence working on the task, perhaps first by themselves, and subsequently in pairs or other small groups. For this stage, the teacher could well anticipate that some students might experience difficulty. We suggest that teachers not try to predict which students might experience difficulty, but allow all students to commence working on the task and decide which students need assistance based on their responses to the task.

Most importantly, following our model of lesson planning, the teacher would have prepared task variations, termed *enabling prompts* as discussed in Chap. 3, that reduce an aspect of the demand of the original task, rather than changing the task altogether or gathering such students together for a teaching session. Examples of enabling prompts that form a bridge to understanding the original task might be to

- Have available some sheets of the required dimensions—even some with the corner squares already marked—and to invite students experiencing difficulties to cut out the squares and make boxes
- Provide a sheet that is already marked in a square grid or
- Have some photos of nets as well as the corresponding folded boxes

The intention is that once students experiencing difficulty have completed the variation(s) on the task, they will be more likely to be able to proceed with the original task as posed and will be able to both follow and make contributions to subsequent classroom discussions.

Of course some students will finish the original task before others. We recommend that teachers seek to extend those students' experience with this task, rather than posing something completely different. For example, students who complete a number of solutions could be asked to

- Search for patterns
- Try to find all possible solutions

- Find different ways to represent their answers or
- Explore closed extensions such as "What are the largest and smallest volumes possible, and how do you know?"

Such task variations are termed *extending prompts*.

The next stage is to allow students, chosen because of the nature of the approach they used, to present their work to the class with questions and comments invited. We agree with Stein, Engle, Smith, and Hughes (2008) that teachers can productively spend much of their time during the student work phase making decisions on which students to invite to comment and the order in which they will be invited.

Finally the teacher summarises the students' insights and strategies, and even adds other suggestions that illustrate themes or commonalities in the students' responses, and compares the outcomes with the goals of the lessons stated earlier.

In terms of the principles of teaching listed in Chap. 2, it is assumed that the teacher would state the mathematical purpose explicitly and describe the ways it is intended the students would work. The open-ended tasks allow teachers to build on what the students know, and can readily be connected to stories. It would be possible, for example, to create a story around the task involving a confectionary manufacturer who wanted to design boxes, and who already has plenty of sheets of the size specified above. The students can explore what might be the best box. The open-endedness also provides challenges as well as allows students to make decisions; it provides teachers with a ready way for them to adapt the task for those students who need support, and to extend those who are ready; and the mode of learning fosters communication and gives students something to share. The teacher can prompt transfer by posing similar tasks subsequently.

Insights into Related Teacher Actions Based on Observations During the Project

Of course, teaching classroom lessons, especially when using unfamiliar formats, is complex and offers a range of challenges. One of the goals of the research project was to illuminate some of those challenges so that teachers can be made more aware of ways of overcoming constraints or difficulties they experience. This section provides some insights into the experience of posing open-ended tasks in classrooms.

Conventional lessons, in which the teacher gives an explanation of a process with the rest of the lesson involving student practice, are easy to design, manage, and pace, and the monitoring of student work is also straightforward. In contrast, lessons based on open-ended tasks are more complex and require a range of teacher actions, some of which may be unfamiliar.

To set the context for the subsequent discussion, the following was a plan for a lesson sequence developed jointly with two project teachers teaching a grade 8 (age 13) class. In the first lesson, a suggestion to the teacher was as follows:

Have a photo of a rectangular prism shape placed where the students can look at it, but not copy directly. Ask them to look, then return to their desk to draw what they saw. Discuss their drawings. Emphasise the key features of effective drawings. Also emphasise the appropriate terminology, such as dimension, length, height, width, rectangular prism...

In the next lesson, it was suggested that the teachers pose this task:

> A set of 36 cubes is arranged to form a rectangular prism. What might the rectangular prism look like?

The teachers were asked to emphasise that there are many different answers, and to have cubes available in case some students needed them. The teachers were also invited to review what students found, especially noting the range of responses, and the commonalities, and to explain that 36 cubic units is the volume. They were then asked to pose the following task:

> A set of 600 cubes is arranged to form a rectangular prism. What might the rectangular prism look like?

In the review, the teachers were asked to emphasise approaches in which students either drew a labelled diagram or just wrote a statement.

Following are some insights about the opportunities and constraints in using such tasks, based on observations of the teaching of this particular sequence:

Finding Out What the Students Are Actually Doing Is Difficult

All teachers make judgements on lesson and student progress based on what they see and hear the students doing. It is possible to gauge the readiness and progress of some students from their comments or questions. The progress of the other students can be assessed by inspection of their work products. It is generally assumed that teachers will examine the work the students are doing and make pedagogical choices based on inferences about the students based on what they see. It seems that this is more difficult than it appears, as the following incidents illustrate.

It seemed that students could visualise the different prisms more easily than they could draw them. Yet it was difficult for the teacher to see the difficulties the students were having with the drawing. For example, some students were able to draw the outline of the shape and then fill in the cubes but were doing this badly. Looking at their final products, the teacher could not interpret the sequence in which their diagrams were drawn.

Teachers need to make particular efforts to determine the processes by which students arrived at their solutions, and to be aware that they need at times to look beyond the obvious.

Sometimes Communicating the Precision of Mathematical Language Is Helpful

Mathematics is a precise language. This creates a need for accuracy in the ways teachers describe ideas to avoid potential confusion by students. The following are two incidents of inexactness in observed lessons.

The first was when the students were asked to replicate the rectangular prisms made from cubes presented on photographs. Neither teacher did this exactly and one even used the incorrect numbers of cubes and none of the students drew her attention to this. It was not immediately obvious that it was wrong, but any thoughtful inspection of the photographs would reveal that the dimensions the teachers were drawing were not correct. Similarly both teachers were loose in their use of language, saying "cube" when they meant "rectangular prism" and sometimes "square" when they meant "cube". There are clearly advantages in modelling the precision of language that is one of the characteristics of learning mathematics. Many of the difficulties that students experience in learning mathematics are associated with the level of their familiarity with the language, and particularly the subtlety of the meaning of some of the terms that they meet. Specific attention to precision of language use by teachers, especially as students progress into secondary school, may reduce some of these difficulties.

It Is Important that Questions Are Purposeful

As indicated earlier, there are advantages in building learning on what the students know. But it is important that teachers gather information from students in an efficient and purposeful manner.

Perhaps when commentators on teaching emphasise not "telling", this has an effect of producing a rather unfortunate pedagogical strategy which is both time consuming and wasteful in terms of directing the learning of the students. There are times when providing information to students is helpful and efficient, but it is no simple task to know when to provide information and when to hold back.

There Is No Need to Talk All the Time

Both teachers who were observed teaching this sequence had a tendency to talk to the class more or less continuously. Most of the time, the students were working on a task when the teachers were talking to the class. This did not seem to be helpful. In one case, there were many process instructions yet it could have been expected that students could sort out for themselves how to go about solving the problem. Indeed the teacher talking seemed to encourage the students to talk, increasing the level of ambient noise and presumably creating distractions for some students. It may also be worth considering that one of the senses is sight, and that teachers encouraging students to watch what they are doing, as distinct from talking to them can be helpful.

It Is Useful to Be Aware of Reluctant Contributors

A surprising observation was that there were students who found suitable solutions, and in some cases a range of solutions, but who did not volunteer their solutions even when it would have meant that the class would finish and they could go to lunch. It is interesting to explore the reason that they seemed reluctant to contribute. Perhaps it was from solidarity with the rest of the class and not wanting to seem as though that they were somehow supporting the teacher. This emphasises that it is sometimes difficult to make judgements based on student responses, and even students who provide no answers may well know more than they are letting on.

The previous discussion indicates some of the challenges that teachers face. Each of these challenges is worthy of further investigation. They also highlight the subtlety and complexity of teaching, especially using non-routine tasks, emphasising that each of the elements of using non-routine tasks can be incorporated productively into the learning of prospective and practising teachers.

Examples of Content-Specific Open-Ended Tasks

Some examples of content-specific open-ended tasks representing a range of levels and content foci are presented below. While these tasks were drawn from our project and so are suitable for students in middle-years' classes, it is straightforward to create similar tasks for students in the early years of schooling, and also for those studying advanced topics in the upper secondary years. The tasks, presented to exemplify this type of task, are as follows:

Find two objects with the same mass but different volumes
Draw some closed shapes with 6 right angles
Draw a line 1 m long on this page
The perimeter of a rectangle is 20 cm. What might be the area?
Draw (on squared paper) as many different triangles as you can with an area of six square units
A number has been rounded off to 5.3. What might be the number?
The median age of people in a family is 18. What might be the ages of the people?
Two fractions add to be ¾. What might be the fractions?
Give 15 numbers between 5.01 and 5.1
A paddock in the shape of an L has an area of 1 hectare? What might be the perimeter of the paddock?

All of these tasks have more than one possible solution, they allow students to make decisions, they can be accessed at various levels, they offer opportunities for extension, and the discussion of the possible solutions creates opportunities to emphasise important mathematical ideas. Further examples of such tasks, with details of their purpose and some information from classroom trials, are presented in Chap. 13.

It is noted that using such tasks may seem unfamiliar to teachers at first and one of the goals of the project was to describe the ways that teachers used the tasks. Some of their reactions are presented in the following section to illustrate not only what is possible but also the ways that teachers interpret and respond to such information.

Some Reactions from Project Teachers to Open-Ended Tasks

The project explored the nature of the learning based on such open-ended tasks, the opportunities that such tasks offer to students, and the constraints that the tasks create for teachers. After the teachers had worked with the respective task types, they completed a survey which asked them questions on these issues. Their unstructured responses were inspected, categorised, and summarised, and are reported in the following section.

The Teachers' Definitions of the Tasks

We were interested to determine how the teachers interpreted the experiences provided by their participation in the project. On a survey, completed after working with such open-ended tasks, the teachers were asked: "If you were talking to a group of teachers about tasks of this type, how would you describe this type of task?"

Nearly all of the teacher responses referred to the possibility of multiple answers using terms such as "multiple answers-multiple methods", "there are a numbers of strategies for finding an answer", and " not only one answer and explore a variety of outcomes".

Many responses also referred to the ways the tasks can be suitable for students of differing readiness, such as "allow students to work at their own level", "use strategies at their own level of understanding", and "access to a range of ability levels".

Various teachers also commented on the emphasis that might be placed on student responses such as "(a need to) focus on sharing strategies", "making generalisations and seeing patterns", and "translating insights into mathematical expressions".

In other words, many teachers were able to restate the purposes and operation of the tasks in the language and form that we had suggested, indicating that at least with the support of associated professional learning, teachers appreciate the intent and value of such tasks.

Some Examples of the Tasks that Teachers Valued

In the survey, the teachers were asked to respond to the prompt: "Of the tasks of this type that you have tried in your class this year, which worked best?" They were also invited to describe the "next two best tasks". Not only did no particular task emerge as more popular, but the most striking feature of the responses was the diversity of tasks that were valued. Examples of tasks that were mentioned a number of times by teachers were as follows:

> Using the map on *google earth*, plan a walk around the school that is 4 km long.
> What might be the missing numbers? _ _ × 8 _ = _ _ 0

These tasks are appropriate exemplars of open-ended tasks in that there is a variety of possible responses to each, the range of responses can be explored by students and teachers, the students have to make choices in finding one or more solutions, and the problems are not solved by the application of a procedure.

There were also examples such as the following suggested:

> How much water is wasted by the school drinking taps over a year?

This has some characteristics of open-ended tasks in that the students have to make active decisions on what is important and how to collect data, and there would be a sense of personal ownership. Interestingly the task also exemplifies the characteristics of contextualised tasks (see Chap. 5) in that it is derived from a practical context.

The teachers' responses indicate that their suggestions of open-ended tasks are compatible with the materials they had been presented with in professional learning sessions, indicating that the terminology and task type are familiar to them, and they can articulate the characteristics of these task.

The Advantages of Open-Ended Tasks as Seen by the Teachers

In teacher learning sessions, we emphasised the following potential advantages of using open-ended tasks:

- There is considerable choice in relation to strategies and solution types; generalised responses and patterns can be found
- There are opportunities for class discussion about the range of approaches used and
- The range of solutions found can lead to an appreciation of their variety and relative efficiencies

After trailing some of these tasks, the teachers were asked

> What do you see as the advantages of using this task type in your teaching?

The most common responses related to the choices that students make about their approach to tasks, such as "how various students go about solving maths problems" and "every student has a chance to solve it in their own way".

Many responses related to the nature of the students' thinking such as "encourages students to broaden their thinking", "creativity", "opens up possibilities", "students think more deeply", and on a slightly different note "encourages students to persist".

Teachers also commented on the ways the tasks can be accessed by all students such as "all achieve some success", "can cater for range of abilities", and "work at their own level".

Having used such tasks in the classrooms, these indicate compatibility with the perspectives that we presented to them. The responses also indicate that the experience of using the tasks was as we had anticipated. Teachers noted the idiosyncratic ways that students responded, and seemed aware that this is helpful for teaching. Teachers also showed awareness of the need to support students individually. In other words, the teachers appreciated the purpose and potential of such tasks.

The Constraints on the Use of Open-Ended Tasks as Seen by the Teachers

In the teacher learning sessions, we discussed the potential constraints posed by such tasks, especially the resistance that some students have to taking the risks that such tasks present (see Desforges & Cockburn, 1987). In the survey, the teachers were asked

> What makes teaching using this task type difficult? What are some challenges in using this type of task?

The most common response related to an issue we had addressed, that is that some students prefer more closed tasks. Teachers comments included "some students are not risk takers", "challenge for the students who want to go straight to an answer", "requires thinking", and "the hard thinking and little direction can be confronting for some kids".

Other aspects of students' response that may be connected to their unfamiliarity of such tasks were "students who don't want to put in any effort", "some find difficulty finding an entry point", "their need for confidence", and "some students don't know where to start".

Some teachers clearly saw such tasks as more difficult, noting that some students might experience difficulties such as "limited mathematical knowledge", and "not all students have the right level of learning".

There were pedagogical aspects mentioned such as "not always sure what maths will come out of it", "correcting the different solutions", "holding back on explanations", and "being ready for what arises".

There were also planning considerations mentioned such as "finding the tasks" and "needs additional resources".

These responses clearly arise from reflection by teachers on the use of such tasks in their own classrooms. It is possible that the constraints might act as a deterrent to the use of such tasks. A significant aspect of our project was to explore the obstacles these constraints represent and to develop ways of working with our teachers to overcome them. Clearly, the teachers were able to anticipate challenges in the use of such tasks and so have greater chance of addressing those challenges.

Sourcing and Creating Content-Specific Open-Ended Tasks

One useful feature of content-specific open-ended tasks is that they can be created by teachers for any topic, for any year level.

There are two methods which teachers have used to construct open-ended tasks. Generally, each method is applicable whatever the topic. Many teachers find that Method 1 works more often. The following presents the methods and some examples of how the methods work. Note that it is often helpful to use a context for the open-ended task that is similar to one that you would use for a closed task. In the following no contexts are presented.

Method 1: Working Backwards

This is a two step process.

Step 1. Write down a typical closed question that might be asked in the upcoming lesson, and also write down the answer to the question.

Step 2. Make up an open-ended task which includes (or addresses) that answer.

For example, a typical perspective question might be

What is a bird's eye view of the building we are in? (Assume that the building has the shape of an L.)

A possible open-ended task could be

What buildings do you know of that have a bird's eye view in the shape of an L?

To give another example, a typical decimals question is

"Round off 1.29 to the nearest tenth" to which the answer is 1.3.

An open-ended task could be

What numbers when rounded off become 1.3?

Another typical question on exchange rates is

If the exchange rate is \$A1 = \$0.98 US, how much would be an item marked at \$US 25 in Australian dollars?
The answer is \$A25.51 (essentially the question is $25 \div 0.98 = 25.51$).

An open-ended task could be

An item marked for sale in \$US costs \$A43.10. What might be the marked price and the exchange rate?

In each case, the open-ended question has a different potential from the corresponding closed question.

Method 2: Adapting a Standard Question

This method also involves a two step process. These are

- Write down a typical question, including the answer
- Adapt it to make it an open-ended task

A key strategy for adapting is to progressively remove some of the question parts. To give an example, a typical question might be

$1.3 + 2.8 = 4.1$

Progressively removing numbers and replacing them with blanks get us to a task like

What might be the missing numbers? _._ + 2._ = 4.1

To give another example, a typical fractions question might be

$$\frac{1}{3}+\frac{1}{4}=\frac{7}{12}$$

Progressively removing numbers and replacing them with blanks get us to a task like

$$\frac{1}{?}+\frac{1}{?}=\frac{?}{?}$$

Note: this task is interesting because of the patterns that emerge in the possible answers.

One or other of these methods works in nearly every possible situation, meaning that teachers can adapt their conventional text book tasks to open-ended if they so wish.

Summary

This chapter outlined the nature of open-ended tasks, and the potential for their use in classrooms, as well as presenting some examples of the tasks, and one example of how the task might be extended to a lesson. It also indicated some challenges of which teachers can be aware when using such tasks, responses of teachers to the tasks, and how the tasks can be created by teachers.

The experience in the project was that teachers appreciated both the opportunities and the challenges in using such tasks, but found that the tasks were both manageable and helpful when used in their classrooms. The tasks clearly create opportunities for learning mathematics.

Chapter 7
Moving from the Task to the Lesson: Pedagogical Practices and Other Issues

We alluded in earlier chapters to the challenges and considerations faced by teachers as they seek to take a task and build a coherent lesson around it. We talked about research on task choice, the role of teacher content and pedagogical content knowledge, the need to consider extending and enabling prompts to use alongside the task, and the subtle but important differences in the use of different kinds of tasks.

Put simply, the task of itself is not sufficient to guarantee student learning, irrespective of its quality. This chapter is an attempt to articulate some key teacher actions which have the potential to maximise learning opportunities presented by a given task. The seven actions which we highlight in this chapter are the following:

- Being clear on the mathematical focus and the goals of the lesson for students.
- Considering the background knowledge which students are likely to bring to the task, how to establish this, and likely responses students will make to the tasks.
- Considering ways in which students who have difficulty making a start on the task and students who solve the task quickly might best be supported.
- Monitoring students' responses to tasks as they work individually or in small groups on the tasks.
- Selecting students who will be invited to share during discussion time.
- Focusing on connections, generalisation, and transfer.
- Considering what the next lesson might look like.

The reader will note the connections between this list and the six principles of mathematics teaching listed in Chap. 2. As described in a processing model suggested by Stein, Grover, and Henningsen (1996) and discussed in detail in Chap. 3, the implementation of a task is affected both by the teacher and by the students. According to their differing goals, beliefs, and attitudes, teachers might implement the same task in different ways. They might organise the class in different settings, and might lower the level of challenge. The students affect the task through their interpretations, which might be different from what the teacher intended and depend on many factors including classroom norms, and their existing knowledge and schemes.

P. Sullivan et al., *Teaching with Tasks for Effective Mathematics Learning*, Mathematics Teacher Education 9, DOI 10.1007/978-1-4614-4681-1_7,

In discussing taking a task and using it to create a worthwhile mathematical experience, we consider a particular open-ended task, *Looking for Three More*, and discuss the process a teacher might pursue in order to maximise genuine learning opportunities for as many students as possible. We should emphasise that our description of the work of teacher and students in relation to this lesson can be considered as an amalgam of lessons we have tried with the task, those we have observed, and those which we wish we had taught!

Looking for Three More: An Open-Ended Task to Challenge and Enhance Students' Understanding of (Mean) Average

> Four people in this room have an average height of 148 cm. You are one of those. Find the other three.

This task is adapted from Sullivan and Lilburn (1997). It is an open-ended task, involving content which has a range of applications, with a number of possible solutions for each student in the class. It seems appropriate for students from grades 5 to 10. It is worthwhile to contrast it with "typical text book treatment". The task in a textbook, at best, might give the average of four numbers (148), and tell students that one of these is 150, inviting them to find the other three. The open-ended task above places the challenge within a meaningful context (height), as well as building in a personal dimension (the individual students' heights), which is likely to increase engagement.

In what follows, we mostly use the term average. Unless stated otherwise, this is taken to be the *mean*.

In the remainder of this chapter, we talk through the actions of a teacher and students prior to, during, and after the lesson, as they attempt to make the most of the opportunities presented by the use of this task. In this discussion, we move backwards and forwards between general principles of turning a task into a lesson and the specifics of how the *Looking for Three More* task played out.[1] The following discussion is a mix of the notes from the classroom observation and commentary written later. These are not distinguished from each other to make the flow of the discussion easier to read.

Being Clear on the Mathematical Focus and the Goals of the Lesson for Students

It may be stating the obvious, but it is important that the teacher goes into the lesson having a clear idea of the mathematical focus of the day. Simon (1995) discussed

[1] Teachers depicted in photographs in this chapter have given permission for these photographs to be used in this publication. In the case of the students, both the students themselves and their parents have also given such permission.

"the creative tension between the teacher's goals with regard to student learning and his responsibility to be sensitive and responsive to the mathematical thinking of the students" (p. 114). Although a feature of high-quality tasks is that they can often take the teacher and students in unanticipated directions, a well-prepared teacher will nevertheless have a likely lesson trajectory in mind as they begin the lesson.

In clarifying the focus, a teacher might ask, "what is it that I want my students to know and be able to do after today's lesson which they maybe did not know and perhaps could not do before the lesson?" Ideally, the answers to these questions emerge as the teacher works individually through the task, and then discusses the task with colleagues.

For the *Looking for Three More* lesson based on this task, some appropriate goals might be as follows:

- Students have both a conceptual and procedural understanding of average. That is, they think of average conceptually, for example, as a kind of balance point if all of the scores were laid out in a line, or as the result of "fair sharing" of scores. In terms of procedure, they know how to calculate average from a set of scores, how to find the total of n scores given the average, and how little adjustments to scores might affect the average.
- Students know their own height in centimetres, and have a sense of typical heights of students in grades 5 and 6.
- Students are aware of the procedures for measuring someone's height to the nearest centimetre.
- Students have a sense of the kinds of sets of four scores which can average 148 cm, and that there are many possible sets of four scores which could achieve this.

Considering the Background Knowledge Which Student Are Likely to Bring to the Task, How to Establish This, and Likely Responses Students Will Make to the Tasks, Including the Difficulties They Might Experience

Although we acknowledge that such preparations for mathematics lessons are difficult to maintain, day after day, there are obvious benefits to teachers being well prepared. Prior to considering the likely responses which students might make to a given task, it is desirable for teachers to have worked through the task *themselves*, possibly discussing their solution(s) with colleagues, before moving to a consideration of typical student responses, including difficulties which might confront them as they work on the tasks. Where possible, there is obvious benefit also in talking with teachers of prior grade levels about the kinds of experiences in which the students have participated previously, to gain a sense of the background experience and knowledge which students are likely to possess. Access to relevant curriculum documents and relevant research papers would all contribute to an ideal preparation.

For *Looking for Three More*, the teacher anticipated that the students would be likely to have little experience with the concept of average or the procedure used to calculate average of a set of numbers. This expectation informed decision making early in the lesson. The teacher also solved the main task individually, and compared notes with colleagues who had also done so.

Considering Ways in Which Students Who Have Difficulty Making a Start on the Problem and Students Who Solve the Problem Quickly Might Best Be Supported

We argued in earlier chapters that the construction of enabling prompts and extending prompts for each of these situations is a desirable part of lesson preparation for the teacher. Sometimes an enabling prompt can be incorporated into the introduction—a kind of "just in case they have no idea" approach to the introduction, which assumes little pre-knowledge on the part of the students, but does not greatly bore students who are well prepared for the task. Even if this occurs, there needs to be other enabling prompts, once the students are underway with the main task. Examples of this are given shortly.

Similarly, there may be students who find the initial task straightforward, and need to be challenged further. Sometimes this challenge can take the form of "joining the teaching team" and supporting other less capable students, but additional challenges are also desirable.

The teacher prepared several *enabling* prompts for those who might have difficulty starting on the main task. These included, "can you please find the average height of your group?" and "how would you find the average of your height and my height?" The potential *extending* prompts are included later, in the discussion of the possible content for the next lesson.

We noted in Chap. 3 the finding from Stein et al. (1996) of the tendency of teachers to reduce the level of potential demand of tasks in the face of student difficulty or anticipated difficulty. The enabling prompts are carefully chosen to provide a means by which students can quickly return to the main task for the day.

Monitoring Students' Responses to Tasks as They Work Individually or in Small Groups on the Tasks

Although the teacher will have hopefully considered how students might respond to the task in advance of the lesson, things rarely go exactly according to plan, and it is important for the teacher to observe the students at work and gain a sense of common strategies and difficulties, if any.

"Who can tell me something they know about average?" The lesson started with an attempt to ascertain the previous experience and level of understanding of this grade 5/6 class about average. The intention was to connect to this experience and understanding in order to further develop their knowledge and skills about average. Most looked fairly blank, but one girl volunteered, "it's not the highest or the lowest, but somewhere in between". There were a couple of grunts of acknowledgement from other students.

"Can you give me an example of where someone might talk about averages?" Sports were mentioned as one context, with average goals kicked by a team or player in football being an example of the use of average. The comment that average was the most common value prompted the teacher to say that this was called the mode and that it was one form of average. The teacher took the chance to mention that the average we would be focusing on today was the mean, and that sometimes in sports averages, the mean could be a score that had never been attained by the individual or team. An example of this would be Ricky Ponting's Test cricket batting average, 55.22, a score he clearly had never made. Another student explained that to find the (mean) average, "you add up all the numbers and divide by how many there are".

It appeared clear at this point that there was no deep knowledge of average among the group, and the usual classroom teacher had confirmed that this was not a topic they had addressed in the 5 months of the year to date.

The teacher explained that the class would be working through two problems that would provide some insights into the notion of (mean) average, one as a whole class, one in small groups.

We see that the teacher has not only spoken to other teachers about the previous experience of the students which may be relevant to this task, but also taken the chance early in the lesson to confirm the sense of what students might know in relation to average.

$15	*$50*	*$5*	*$30*

Four volunteers sat on the carpet and were given plastic pockets with differing amounts of money (totals of $15, $50, $5, and $30 respectively). The task was posed to the class: "If this money was shared equally between our four volunteers, what would happen, and how do you know?" He had anticipated that students might say, "Julian would get more, Shalinika would get less, …", but there was no response for a period of about 20 s. Then someone said, "they would get $25 each, because there's $100 altogether, and I know that 4 times 25 is 100".

Several others nodded, and the teacher then indicated that this was correct, and therefore "Julian, who had $15 would get more; Shalinika who had $50 would get less", and so on.

The teacher explained that finding the (mean) average of a set of numbers (or dollar amounts) was like "equal sharing", and that when we did add all the numbers together and divide by how many numbers there were, the result was like a fair share.

It was clear that this introductory discussion of averages in the context of money was important if students were to be in a position to work on the *Looking for Three More* task, given their apparently "shaky" understanding of the concept and procedures involved in average.

The teacher then indicated to the class that they would now move on to the main task for the lesson.

A brief introduction of the kind above served at least two purposes: it enabled the teacher to see what the students already knew about average (or at least some of them); and it provided some teaching of the notion of average which was likely to scaffold students into the main task for the lesson.

The teacher asked how many of the students already knew their height in centimetres. The only two to respond were two "early birds" who had measured their height just prior to the lesson, when the measuring process was being set up. It is interesting to note how few students in the middle years of schooling, and in our experience teachers, know their height in centimetres.

The teacher moved around the room while speaking, indicating that students would be working in groups of three based on where they were sitting, and placed a sticker on one student in each group, careful to have some of the taller and shorter students in the collection of students who were given stickers. The problem was then posed:

> Four people in this room have an average height of 148 cm. You are one of them. Find the other three.

The teacher was still a little uncertain as to whether 148 was the best figure to use, and by using the word "people" in the task prompt rather than "students", it meant that teachers could be included in the four if necessary (there were four teachers observing the lesson at the time).

It was explained that each group of three had to work together to find the other three people to complement the "sticker person", thus creating a group of four with a mean height of 148 cm. "So Ben, your group needs to help you find three others to make your group of four. Some of the others you need might be already in your group, but it is likely that you will need to do some searching".

The teacher's recent experience in middle years' classrooms indicated that relatively few students knew their height in centimetres. Tape measures had been taped to the wall therefore in three places around the room.

Some were two tapes joined together, such that a student with little familiarity with measurement might read their own height as 36 cm rather than 136 cm, and this happened initially, with some necessary correction, as students worked together. Students started to measure the heights of each member of their group. There were a number of teachers in the room, and it was necessary for several groups to have the problem clarified by the teachers. For example, there was initial confusion about whether they needed to measure beyond their small group. The general pattern was that the grade 5/6 students were busy caught up in the fun of measuring each other, putting the actual problem aside for the time being.

There were several groups who had taken the task to find four people who all had a height of 148 cm. Certainly, this would be a solution if they could find them, but the mathematical purpose was still lacking at this stage for these students.

While most students' books listed some names and heights, two of the students' books looked like this:

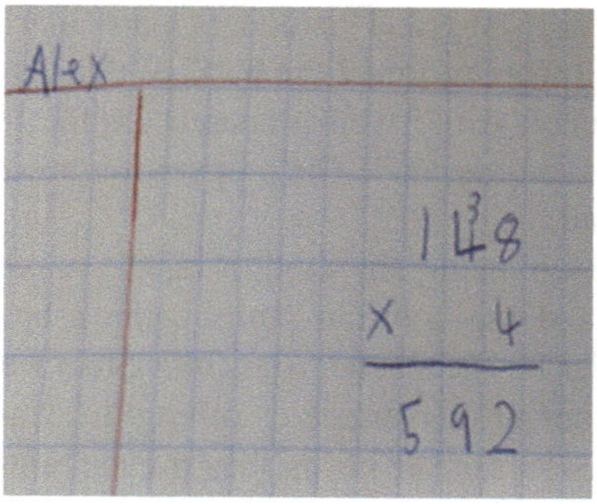

In the case of Alex's book, it was likely that he had built upon the work with the money or his prior knowledge in finding the total height of the four students who would form his group (592 cm).

In the case of Samantha's work, she had totalled four students' heights (giving 593 cm), and calculated the average height as 148.25 cm, just a little more than the desired target.

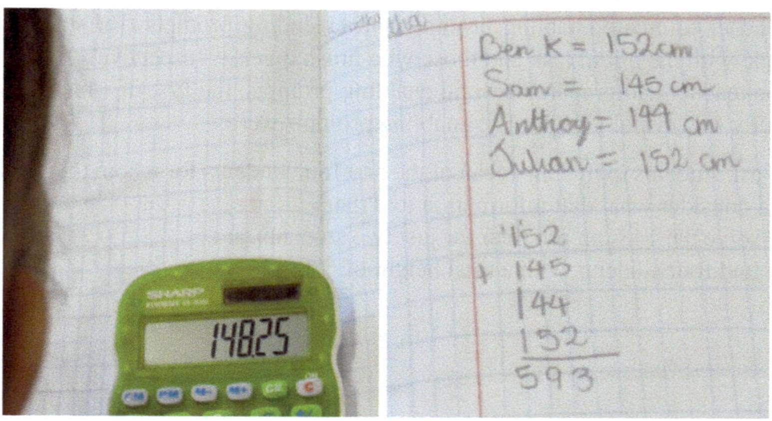

Samantha's group was engrossed in a conversation about how to deal with the "extra 0.25 cm". One view was that they needed to find four people, each one centimetre shorter than the previous four. A second view was that they needed to find four people, each 0.25 cm shorter than the previous four. The teacher simply asked, "when you find the people you need, what will the total of their heights be?" and walked away. The teacher appears to have made the decision to just "sow a seed" through the question, assuming that the students will then discuss it, and take appropriate action.

Interestingly, an open response item addressing similar content was on the grade 9 (calculator allowed) Australian national mathematics assessment in 2009.

The wording was as follows:

After 15 games, Sam's average (mean) number of points per game was 17. After 20 games, his average had increased to 21. What was his average number of points per game in the last five games?

The percentage success rate for Victorian students was only 10%. Clearly, this content is challenging for even grade 9 students, particularly if they have not experienced tasks of the kind which formed the basis of this lesson.

We see that the teacher, while well prepared for the lesson, was nevertheless making many decisions "on the run". The preparation however enabled these decisions to be well informed.

One of the challenges in using any kind of task is maximising the chances that other students understand the solution paths offered by individuals during the "reporting time". In order for the teacher to facilitate this well, it is important for the teacher to understand the various strategies offered by students. A teacher is well prepared for the discussion time if he/she has a clear sense, during individual or small group work, of the kinds of strategies which students are proposing and using.

It can be important during this monitoring phase that the teacher, for some time at least, stands back and observes and listens, without intervening. Lampert (2001), in describing a lesson she taught in great detail, noted that she spent about 20 minutes out of the 30-minute small-group part of the lesson just watching and listening, and the rest of the time interacting.

We have discussed the desirability in preparing to use tasks in the classroom for the teacher to create enabling prompts and extending prompts (for students who struggle with the initial task and those who finish quickly, respectively).

The teacher had prepared several enabling prompts, the first of which was used helpfully for several groups. The enabling prompts were:

- To ask students to find the mean of the first four students for whom they had data, and consider what that information told them
- To solve the 148-cm problem for just two students and
- To find four students with a total height of 600 cm

The three extending prompts were

- To find a second set of students (all different from the first) with a mean of 148 cm
- To suggest and carry out a plan to find four students who had a mean now of 149 cm, measuring as few additional students as possible and/or
- To find a set of four students with a mean of 148 cm, but including three students with heights greater than 148 cm, or three students with heights below 148 cm in the set

None of the extending prompts was used, as the teacher found that the main task was sufficiently challenging for all students.

Selecting Students Who Will Be Invited to Share During the Discussion Time

In contrast to having everyone who wants to contribute having the "floor", the monitoring above enables the teacher to select carefully those students whose sharing will provide an opportunity to maximise the learning of the whole group. Stein, Engle, Smith, and Hughes (2008) emphasised the importance of "selecting particular students to present their mathematical responses during the discuss-and-summarize phase, ... purposefully sequencing the student responses that will be displayed" (p. 321). One possible ordering of reporting back is first to have a student or group share that made some progress but did not completely solve the problem, possibly revealing a common misconception or difficulty. This could be followed by a student or group that solved the problem in a satisfactory but common way. Finally, a student or group of students could present who provided an innovative and/or particularly elegant solution. It is likely that it will be sufficient for two or three students or groups to share.

After about 30 minutes of measuring, calculating, and additional measuring, most groups were making progress, and the students were called together. Two sets of data (the four names and measurements in each case) were recorded on the board, as follows. Samantha's group had not solved their problem at the time when the class was called together, though most other groups claimed to have a solution with which they were happy.

As the representative of the first group chosen by the teacher to report, Samantha talked through the process they had undertaken, and then explained their difficulty (responding to the total of 593 cm), and other students were able to explain that they just needed to replace one person on their list with someone 1 cm shorter, to bring the total down to 592 cm.

Other students talked through the different strategies they had used in arriving at their solution (or at least in their progress to date). They were quite frank about their early difficulties before they had made progress. The teacher "stood back", encouraging students to talk to each other during this phase.

The teacher had placed a strip of masking tape horizontally on the whiteboard at 148 cm above the floor. He invited the class to consider what we would expect to see when a group of four with the appropriate mean lined up. There were three kinds of responses, all indicating a different understanding of mean:

- All four would be right on the line.
- All four would be close to the line.
- The four would kind of "balance out" with some above and some below.

The teacher posed the question, "Could we have three above and one below, and still have an average of 148 cm?" Most felt this was not possible.

The students came out in groups of four and lined up, and the class was invited to make a judgement as to whether each respective set could be correct. There was a nice mixture of fours, with students well above, just above, on the line, just below, and well below the line.

For the teacher, this visual component of the lesson was the highlight, and was hopefully a major contributor to students' conceptual understanding of average. If time permitted, or in another lesson, the teacher could also raise the issue of variance, noting that the sum of the "bits above the line" have to add to the sum of the "bits below".

Focusing on Connections, Generalisation, and Transfer

Depending upon the task and the teacher's purpose for it, the final phase provides the chance to encourage students to think about making connections between student solutions or connections with previous work. Another focus might be generalising from what they have learned and/or what can be transferred to new tasks, with questions like, "what kinds of things have we learned today which will help us to solve other problems?" "Could that method or strategy work no matter what numbers were involved? When would you use that particular strategy other than in tasks of this kind?"

Boaler (1993) noted that it seemed likely that an activity which engages students and enables them to attain some personal meaning would enhance transfer to the extent that it allows a deeper understanding of the mathematics involved. Boaler believes that the key is to analyse mathematical situations and understand which aspects are central.

The teacher took the chance to summarise what has been learned, with statements like the following:

- We have learned that there are several different ways of thinking about average, and that the most widely used one is called the mean.
- We have seen that a number of different scores or measurements can all have the same average.
- We have seen that the average gives us a kind of "balance point", and that scores contributing to an average can be mostly above the average, mostly below, or a mixture of both.
- We have seen that we can adjust the average by changing a single score, because a single score has an influence on the mean.

For students in grades 9 and 10, the two formulae $\left(\sum x / n = \overline{x} \text{ and } \sum (x - \overline{x}) = 0\right)$ could be presented, to introduce more formal representation of what has been learned.

On some occasions, the teacher might ask a student to summarise what has been learned, but in this case, the teacher made the call that the students had been given an opportunity to contribute and the summary was best to come from the teacher.

Considering What the Next Lesson Might Look Like

When we work with teachers in schools in professional learning programmes, we often ask, "What would the next lesson look like?" This is an important question, as often teachers' responses imply that the topic addressed by the particular task of the day is now "covered" and we can move on. As argued in Chap. 2, if we are to build fluency, it will be important to think of other tasks which can reinforce what was learned from today.

For the *Find the Other Three* lesson, it is relatively straightforward to create follow-up tasks, in which students calculate (mean) averages in a variety of contexts, adjusting them as necessary to change the mean as required. Sometimes these might correspond to the pre-planned extending prompts, but they also might be quite different.

Sample tasks could be:

- Five people tried the long jump and their average jump was 2.70 m. What might their jumps have been?
- Now change just one of your jumps to make the average 2.80 m.
- The fishing problem (see Chap. 8) or a variation which adds median to mean: "Seven people went fishing and the mean number of fish they caught was 4 and the median number was 6. How many fish might they have each caught?"
- Imagine that the world's tallest man (Bao Xishun) walked in the door (236 cm). Could he possibly belong to any groups of four from the previous day? How about the world's tallest woman (Yao Defen, 233 cm)?
- Let's develop a spreadsheet with all of the class data, which presents all the possible combinations of four students, with the relevant average heights. (There would actually be 12,650 different groups of four, in a class of 25 students.)

While it is unlikely the teacher would use all of these, it is possible of course to offer students a choice among these, with pairs of students (say) each picking one of interest to investigate.

Differences Between Task Types in Relation to the Process of Turning a Task into a Lesson

There is an issue that has not been discussed in this chapter to date, and is mentioned briefly now. How might the process of turning the task into the lesson vary according to the task type? This can be considered one by one as follows:

Purposeful, representational tasks. Lessons using these types of tasks are less likely to involve a motivational "hook" at the start. The mathematics and solution paths are also likely to be narrower, with the mathematical focus clear. If built around a game, the instructions will be given, possibly with a demonstration game for the whole class to observe. There is also less likelihood of the lesson taking a major detour along the way than for other task types. The reflection on the lesson is likely to take the form of "if you played this game or did a task of this kind again, what would you do differently?"

Contextualised tasks. Tasks of this kind will usually start with a discussion of the context, and students' experiences with it or awareness of it. ("Did anyone see that story on the news about …?" "How many of you play soccer?" "Have you ever been travelling and seen one of those signposts that shows how many kilometres to 9 or 10 places in different countries in all different directions?"). The lessons will also generally finish with a return to the context, as well as an overview of the mathematical goal for the lesson.

Open-ended tasks. These tasks are generally of a form that most students can make a start, provided these kinds of tasks are not new to them. Some of the challenges will be to have enabling and extending prompts for those who need them, and to convince students that they should challenge themselves to move towards generalisation as far as they can. It has been claimed that all mathematics lessons should lead towards generalisation, but particularly so for open-ended tasks, in our opinion. The lesson review will be important so that the teacher can draw together the mathematical goals for the lesson, given the range of responses which students usually provide to open-ended tasks.

Summary

Using a specific open-ended lesson, drawn from several observed examples of the use of the task which underpinned it, we discussed a major role of the teacher of mathematics—to choose meaningful and relevant tasks, and develop them into worthwhile mathematics lessons. Early in the chapter, we noted that, however good it is, the task is not enough.

We have seen that there are advantages in thinking through carefully before the lesson the mathematical focus, the goals for students, how they might achieve these, and possible enabling and extending prompts. As the lesson proceeds, the teacher needs to probe students' thinking without extensive "telling", and gather a sense of how individuals and groups are responding to the task(s). In a review of the lesson, the teacher has the chance to select carefully those students who might share and then summarise what has been learned mathematically, and about the context, if any.

Of course, we are not claiming that the seven actions which form the basis of this chapter would all play out in all lessons or even in the same way. Other teachers might use the task "Looking for Three More" in quite different ways, and yet still be highly effective. The work of Shimizu, Kaur, Huang, and Clarke (2010) provides ample evidence of the different classroom norms and teacher actions evident in different countries. It would be foolish to claim that there is one recipe in moving from the task to the lesson.

However, what is clear from our study and the work of other scholars is that the teacher's role, carried out effectively, involves many dimensions, much preparation, and a lot of thinking on the run. Such actions could profitably be the focus of teacher professional learning.

Chapter 8
Constructing a Sequence of Lessons

While tasks are the basic building block, teachers need to create lessons to ensure that the potential of tasks are realised, and also need to plan sequences of such lessons to build coherent learning experiences for students. The following presents some details of the development process and tasks used in one of the sequences created by project teachers. It includes information about the tasks and the instructions given to other teachers and some comments on the intent of the tasks. Some items from the suggested assessment instrument are presented as examples of the type of observation notes that were gathered.

Introduction

As well as examining various aspects of task use, as is described in the other chapters in this book, the project also sought to study the ways that teachers constructed, used, and evaluated sequences of lessons with related foci, described in some places as units of work. The project studied in detail four such sequences, three at grades 5 and 6, and one at grade 8. In each case, the participating teachers were invited to define the focus of the sequence and to gather the usual resources they used in planning. The project team met with the teachers to clarify the focus of the particular lesson sequence, and then suggested various tasks, representing each of the types discussed in Chaps. 4–6, from which the teachers could choose in planning to supplement their own resources.

The project team then studied the teaching of the sequences. This involved interviews with the teachers before each lesson; structured observations of lessons including interactions with students, usually with two observers per class who made notes and took photographs; and interviews with the teachers after each observation. There was some data collection from the students at the end of the sequence.

We were surprised at the diversity of types of sequences that the groups of teachers prepared. One of the grade 5/6 sequences was based around a contextualised

P. Sullivan et al., *Teaching with Tasks for Effective Mathematics Learning*, Mathematics Teacher Education 9, DOI 10.1007/978-1-4614-4681-1_8,

theme, another was designed to be highly individualised, and the other two were of the form we had anticipated. One of these sequences is elaborated below.

It should be noted that the respective teams of teachers had volunteered to create these lesson sequences and so may well be more confident and adventurous than is usual. Nevertheless our experience suggests that the creation of lesson sequences is highly individualistic, and so the following sequence is not presented as exemplary or even representative, but to indicate what is possible and to illustrate how consideration of sequences can enhance our understanding of the ways that particular tasks contribute to learning.

A Sequence of Lessons on Presenting and Interpreting Data

The school in which this sequence was developed serves a middle-class community on the outskirts of Melbourne. Teachers Ms. Y (grade 5) and Ms. Z (grade 6) were active members of the TTML research project and familiar with definitions of the task types.

The school supported the teachers' participation including by releasing them for a full day's planning of the lesson sequence. Before the planning day conducted away from the school, Ms. Z and Ms. Y had decided that the focus of the lesson sequence should be on "Interpreting and Representing Data". They reported that in previous years, they had asked their students to collect data from which they had learned how to make pie graphs and bar charts. The teachers' stated intention was to teach a lesson sequence that would also incorporate the analysis of data and interpretation of graphic representations.

Ms. Z and Ms. Y described their aims as being to:

- Challenge students
- Provide students with engaging tasks
- Use each of the TTML task types and
- Include a project incorporating all of the task types, over the duration of the sequence, incorporating concepts on which the children had been working, drawn from the theme of environmental sustainability

As most lessons were 90 min long, the teachers stated that they preferred tasks that involved students moving between small groups and the whole class as a way of keeping them engaged and motivated. The teachers also wanted to ensure the children had adequate time for reflection and review at the end of each lesson.

Ms. Z and Ms. Y noted that the design of learning experiences was constrained by the fact that while students had daily access to four computers, they also had weekly access to a computer laboratory. A further constraint was that the lesson sequence would be taught to all grade 5 and 6 students (there were three grade 5 classes and three grade 6 classes) so they were conscious of the need to adapt tasks to these differing grade levels. They were also aware that there was a need to use newly created tasks since some of Ms. Z's grade 6 students had been taught by Ms. Y the previous year.

On the planning day, initially the teachers worked together, building on one another's ideas. Their initial preference was to devise a project that would be conducted progressively over the entire lesson sequence. Building on a current theme of the environment, they termed this "the Weather Project". The intention was that students would collect temperature and rainfall data at the school and compare it with the inner-city data given in the daily newspaper. By the time they had completed the 12 lessons in the sequence, which would take place over 8 school days during a 2-week period, they would have collected sufficient data on temperature and rainfall to represent and analyse using their newly gained skills and understandings. The goal was that this project would provide meaningful connections between this data representation unit, not only to other aspects of the curriculum but also to students' lives.

The next step was that the teachers planned an initial assessment task to determine how much students already knew about graphs. After considering a range of possibilities from their collection of well-thumbed teacher resource books, they opted for an activity (the game "Hangman") rather than a test to assess students' knowledge. This is described further below.

Ms. Z and Ms. Y listed a selection of possible activities that they had used previously, as well as examining possible sample activities from textbooks and other resources. They then met with the project team who suggested further examples of each of the task types. By mid-afternoon, they had constructed a lesson sequence consisting of 14 tasks, with 4, 6, and 4 of each of the task types (representational, contextualised, and open-ended), respectively, with a list of alternative tasks for use in future years. Their completed plan included statements of each lesson's mathematical focus, followed by descriptions of a short introductory task and the main task, and questions for whole-class reflection. The details of the lessons are presented in the following.

The teachers also planned specific advice related to pedagogy and some of these suggestions are also presented. The intent in the following is to communicate both the teachers' intentions at the planning phase and some comments on the ways that the tasks were used. The student project that required ongoing collection of data which spanned the lesson sequence is described first, followed by the individual lessons.

An Investigative Project as an Overarching Theme

Initially, the teachers introduced the lesson sequence and explained the weather project in which each mathematics lesson began by recording the previous day's temperature and rainfall data in the centre of Melbourne, along with data gathered from the school weather station (which is some 40 km to the East of Melbourne).

Information on the task that was written as a result of the observations of the project is as follows:

> Students gathered temperature and rainfall data over a period of two weeks. These data were summarised and presented in differing graphical representations, both hand-drawn and computer generated. The students calculated averages and compared local data with the Bureau of Meteorology data for Melbourne.

Students were asked to consider the following rainfall questions:

- How much rain fell in a 2-week period in Melbourne and at our school?
- How much more rain fell in the second week than first week?
- What is the wettest day on the graph?
- Which location had the most rainfall?
- What was the range, mean, mode, and median for both locations?

Students were also asked to consider similar questions for the interpretation of the temperature data.

- What is the least and greatest difference between temperatures on 1 day? In 2 weeks?
- What was the range, mean, mode, and median for the maximum and minimum temperature?
- What general impressions can we draw from the data?

In retrospect, this project seemed to provide coherence to the overall sequence, it utilised a meaningful context to which the students could relate, it allowed time-efficient student involvement in data collection, and it created opportunities for cross-curricula considerations.

Along this recurring theme, there were also purpose-designed lessons. The following are the eight planned lessons in this sequence, noting that each was intended to span around 90 min, and so included more than one task in some cases.

Lesson One: Most Popular Letter

The main purpose of this task was to allow teachers to assess how students represented data that they had collected.

The class first played a whole-class game of Hangman, the intention of which was to allow consideration of the variable frequency of letters and winning strategies associated with such a game. This was followed by a discussion in which students were invited to predict the most frequently used letters which might help them play more efficiently.

Next, students, in pairs, were given some text extracts from magazines, from newsletters, from a text book, and so on. Students then recorded the frequency of each letter in 100 words of their text, entered these data in a table, and then presented this as a graph. The class then came together to record their combined information and test their predictions of the most frequently used letters.

The grade 5s, who had a double lesson, used Excel to create a spreadsheet and input data on the frequency of each letter. The teacher showed them how to use the chart wizard to generate different types of graphs and they experimented with this before coming together to reflect on which kinds of graph are most effective for various kinds of data.

The task allowed the teachers to see how the students chose to record the tally of the letter frequency, then to observe how they chose to represent the collated results, and finally to evaluate the ways that the students reported on their findings. It served as a useful initial assessment of student readiness for the content of the following lessons.

Lesson Two: Introducing Graphical Representations

There were two separate activities in this lesson both focusing on key aspects of data representation: the first emphasised conventional aspects such as labels and the second was about connecting related representational forms.

Features of Graphical Representations

Students were presented with a column graph, where most columns were of varying height, but two were of identical height. The columns were deliberately not in ascending or descending order. No numbers, labels, or titles were put on the graph. The students were asked: "What might this be the graph of?"

Students were asked to propose at least three different sets of data that this graph could represent, and to add whatever information was necessary to complete the graph. They were also told that there were many possible solutions to this problem.

This activity sought to explore whether students could apply prior knowledge about graphs and data to an open-ended task, and encouraged learning about the suitability of different graphs to different types of data.

The teacher led a whole-class discussion of how graphs were used (e.g. to represent sports results). She gave out examples of unlabelled graphs, asking students to decide what this graph might represent, to label it, and to write down three things it indicated. Students shared their ideas.

This activity had the effect of allowing students to make choices and decisions, and in some cases to be creative in their interpretations. These are key elements in the representation of data.

Matching Graphical Representations

The second task in this lesson, matching graphical representations, was adapted from Swan (a more detailed discussion of the task is found in Chap. 13), which consists of sets of 12 frequency bar charts and 12 matching pie charts that were copied onto separate small cards. Figure 8.1 is one example of each of these:

Swan's task also includes box plots, but these were not used in this lesson.

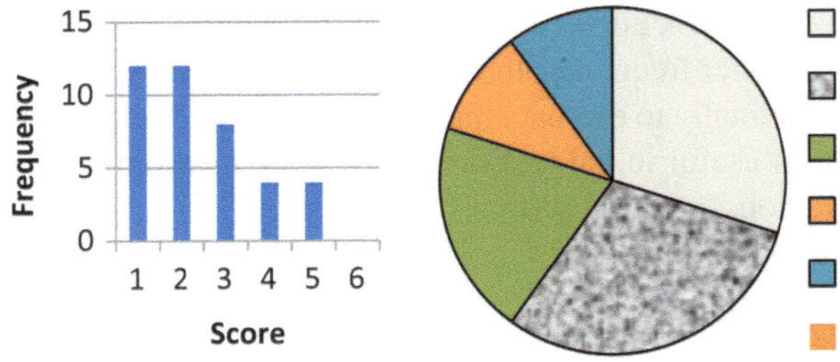

Fig. 8.1 Examples of data representations for students to match

The intention is that students match pie charts and bar graphs that represent the same data, and then compare their responses to those of other groups. Swan (no date) provides teachers with detailed instructions on how to use the cards. For example, the following is the way he suggests introducing students to a bar chart which is drawn on the board:

> This bar chart represents the scores that were obtained when a number of people entered a penalty-taking competition. Each person was allowed six penalty kicks.
> How many people entered the competition?
> How can you tell?
> What proportion of the people scored 1 penalty?
> What is that as a percentage?
> What proportion scored 3 penalties? 6 penalties?

A similar introduction is offered to the pie chart, and suggestions are presented of ways of using the sets of cards and for reviewing the task subsequently.

The purpose of this task is that students have the opportunity to compare and contrast representations, and see the importance of reading the various pieces of information that are provided on a graph. It is a low-risk activity in that it only involves moving cards, and errors can be easily addressed.

Lesson Three: Conducting and Representing Survey Results

This lesson builds on what students know about surveys, and extends this knowledge to creating pie charts. The instructions to the other teachers in the planning team for the first step are presented in the plan as follows:

> Decide on a question that students can survey the class about. Conduct the survey. Record each response as a fraction of a whole.

The next step is taken from the *This Goes with This Lesson* (Lovitt & Clarke, 1988). The instructions were presented to the teachers as follows:

> Students explore transferring data from strip graphs to pie graphs. Introduce the students to creating a 1 m long strip graph from classroom generated data, such as 'Favourite confectionery'. Students divide the strip evenly to represent each child in the class and their preference. Demonstrate how to wrap and fasten the strip into a circular shape to create a pie graph. Each entry represents a fraction of the whole. The students draw and label the graph as fractions, converting to percentages if possible.

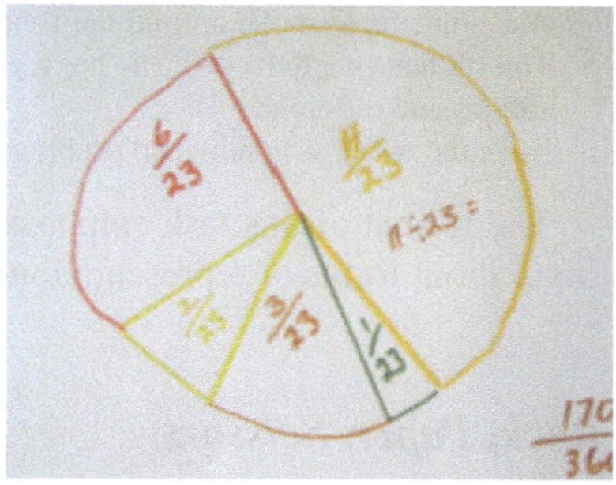

Fig. 8.2 A student pie chart

Figure 8.2 is an example of a student-created pie chart using this method:

This task makes an important contribution to the sequence. Pie charts are in common usage in the media and elsewhere, and they are easier to interpret than they are to construct. This method illustrates a process for constructing pie charts, and so provides a suitable basis for students later to use Excel, or similar, to create pie charts from data they have collected.

Lesson Four: Making Decisions on Representations

There were two complementary aspects to lesson four: the first was about comparing and contrasting graph types; and the second emphasised the need for students to make active decisions about appropriate forms of graphical representation.

Finding Similar Graphs

In this first task in the lesson, each student was given a different graph, sourced from a newspaper, magazine, or the Internet, and had to find other students in the classroom with a similar graph and note the similarities and differences and the kind of information represented. They labelled their graph and wrote down three things that it showed and they then shared their ideas. The instructions to other teachers in their planning team were as follows:

> Introduce a variety of graphs to the students on an overhead projector or smart board, e.g. bar, column, pie, line, scatter plot. Teacher and students discuss the features of these graphs and the types of data they illustrate.
> Give each student a different graph from the media or classroom generated to examine. Allow a few minutes for students to individually examine the graph and:
> Identify the type of graph
> The features of the graph
> The information presented in the graph and
> Alternative graphs that might represent these data.

> Ask students to move around the room and seek other students who have similar graphs. The students are to record who has a graph similar to their own and the ways the graphs are similar and different.
> Rove the room, watching and probing students' understanding of graphs.

The format of this task emphasised for students that they needed to make decisions about forms of representation.

Rock, Paper, Scissors

In the second task in this lesson, following a class discussion, students played the game Rock Paper Scissors in pairs 100 times, recording the results and presenting them as a graph. It is a popular game which students play in pairs and form their hands into one of the three arrangements (rock or paper or scissors), and eliminate each other depending upon choices (e.g. rock blunts scissors, paper covers rock, scissors cut paper). The students were posed the task to find which of rock, paper, or scissors was most likely to win. Students were free to choose both the method of recording results and the type of graph. At the end of the lesson, a whole-class discussion considered the relative effectiveness of different recording methods and graphs. The instructions were presented as follows:

> Students are to explore the most likely winning gesture in the game 'Rock–Paper–Scissors'.
> First pose the question:
> How could we work out the best way to win when playing rock paper scissors?
> Students discuss with their partner what data they might collect. Following a class discussion, the students play the game and collect and record the data they suggested.
> In pairs, students play 100 games and record the winning gestures for each game. Students transfer data to a graph. Students select their own method for recording and presenting the data.
> Students come together with the class to reveal and discuss their results. Which gesture was more likely to arise? In theory, the gestures have an even chance, but what is altering that chance? Why are there differences between the groups? What possible conclusions can we draw from the data?
> Students share the methods used to record and present the data. Students discuss the advantages and disadvantages of representing data in the various ways presented by the class.

This is quite a challenging task but the game is familiar and it does not matter if students first choose an inefficient or ineffective method for recording the results of their games. Again it exemplifies the connection between the data and the forms of representation.

Lesson Five: Two-Way Tables

The purpose of this task was to introduce students to an alternate way of presenting categorical data, especially when there are two variables. An example, drawn from the lesson, of a two-way table is as follows:

> The teacher had asked students the following questions "Do you like Collingwood (a local football team)?" and "Is chocolate your favourite sweet?" The students wrote their

answers (yes or no) to each of the questions on a post-it and placed theirs in the relevant square of the table. The resulting tabulation of the data might be:

	I like Collingwood	I do not like Collingwood
Chocolate is my favourite	7	8
Chocolate is not my favourite	4	4

These data were analysed to create generalisations about the class. Working in pairs, students then created their own two "yes/no" questions. They shared ways of collecting the data, and then surveyed their classmates and constructed a two-way table. The class gathered to share, display, and discuss the data collected, alternative ways to present them, and generalisations that could be drawn from their tables. The instructions to the other teachers were as follows:

Ask students to explain what the table is telling us.
When might it be useful to present information in this way? Who might use it?
Ask several students to share the tables they have created. What conclusions can you draw? What conclusions can you not draw? Highlight the concept of sampling.

This activity illustrates important ways of representing relationships between categorical data. The importance of ensuring that one answer is not just a subset of another was also evident in the student-generated questions.

Lesson Six: Average Height of Class

An additional perspective of representing data is the use of measures of central tendency (mean, median, and mode) and spread (the range). This lesson explored the notion of average (mean) as one of these measures. The context was the average height of students in the class.

In order to find the average height of the class, students brainstormed definitions of average (mean, median, and mode) and methods for collecting data. In pairs, students measured their heights and recorded these on paper streamers. These were laid on the floor in order of height and the teacher used them to model different meanings of average, for example, demonstrating the mean average of two streamers by taking a piece from the longer streamer and adding it to the shorter one to create two streamers of equal length. The class then transferred the data to the board and students worked in pairs to find the mean, median, and mode of the height of the class before gathering to share their methods and results. The grade 6 teacher showed her class how to use Excel to find the mean, median, and mode.

The description of the lesson prepared for the other teachers in the planning team was as follows:

Explain that in this task the students are going to find out the average height of their class. Ask 'What does average mean?' The students brainstorm definitions of average. Students may consult a mathematics dictionary. At the end of the discussion definitions of mean, median and mode are displayed (or written into students' journals).

In the case of mean of two streamers, demonstrate how you might take a piece off the end of a longer streamer and add it to a shorter streamer to create two streamers of the same length, thus finding the mean average of two streamers/heights.
Pose the question 'How might we find out the "average" height of our class?' Students brainstorm methods for collecting data to answer this question effectively.
In pairs, students use a tape measure and streamer to measure the height of their partner. Students may either measure using the tape measure and cut the right length from the streamer or match the streamer to the height of the student and measure the streamer length afterwards. The height and name of the student is recorded on the streamer.
In a large open area, the students organise the streamers to be laid on the floor in order of height.
Back in the classroom, these student data are transferred to the board in descending order.
Ask the students to work in pairs or a small group to find the mean, median and mode of the height of the class. The students may use calculators if required.
Have the whole class come together to discuss their findings. The students share their results and the methods used to find the average with the class. This may be an opportunity to introduce the term 'range'. The teacher asks the students to consider, 'Why are there differences in the results? Which result would they use to describe the average height of their class? Why?'
At the conclusion of the lesson, demonstrate how to find the mean, median and mode in Excel through the use of formulas built-in to the program.

This lesson seems an appropriate way to introduce students to the concept of mean. The streamers provide a visual or concrete model that is easily managed, with the meaning of the mean being obvious in the simple case.

Lesson Seven: Mean Average

Lesson seven extends the students' experience with the mean through two tasks: one task requires construction of sentences, and the other measuring and associated calculations.

A Sentence with Five Words

Students explored the terms mean and average through the following open-ended question:

> A person wrote a sentence with 5 words. The average number of letters in each word was 4. None of the words had exactly 4 letters. What might the sentence have been?

Students explored their own strategies for tackling this task and then shared their responses and strategies. For example, one of the students suggested "Bubbles float on hot air".

This task is interesting for students in that it requires them to be creative, to link different aspects of the curriculum together, and to explore possible meanings of the word mean.

Average Height of the School

In the second task in this lesson, the students were asked to find the average height of students in the school. After revising the concept of average and predicting the average height of the school, students worked in groups of three to discuss their strategy and sample size. They shared these before setting off to collect their data from other classrooms. On returning, each group calculated the average and then shared with the class their method, choice of average (mean, median, or mode), and final result. The teacher prompted them to consider why their results differed, and the comparative accuracy of different methods.

Lesson Eight: Comparing and Contrasting Measures of Centre and Spread

There were two complementary experiences to this lesson, both of which require students to contrast the various terms describing central tendency and spread that they had learned.

Seven People Went Fishing

After revising definitions of mean, median, and mode, students worked individually on the following problem:

> Seven people went fishing. The mean number of fish caught was 5, the median was 4 and the mode was 3. How many fish did each person catch?

After working on the problem, they shared their approaches and the teacher used incorrect solutions to highlight misunderstandings of the concepts.

The instructions to other teachers in the planning team were presented as follows:

> The students may require clarification of the terminology of mean, median and mode before commencing the task. Encourage the students to think of ways to remember each term, such as that the median strip is in the middle of the road.
> Encourage students to work first on the problem individually. Emphasise that there is more than one possible answer to this question.
> After they have been working, draw the students together to share their approaches and responses to the question. Some students may start by calculating the total number of fish caught (35) from the mean (5), i.e., 5 (mean)×7 (people fishing)=35 fish. Other students may created 7 spaces in a row, and enter the 4 (median) in the middle place (4th) of the 7 spaces.
> A third possible approach may see students commence with placing two or three 3s for the mode along their line of 7 spaces.

Clues on Cards

The second task in this lesson required students to work collaboratively in small groups, reconstructing data from a set of cards with clues, such as "The mean of the scores is 5", and "The range of the scores is 10", based on use of the terms mean, mode, median, range, and frequency. The instructions to other teachers in the planning team were as follows:

> Place children in groups of 6. Each child is given a clue card that contains a piece of information that assists the group in reconstructing the data.
> Before reconstructing the data, the students break off into 'expert' groups with students from other groups who have a matching symbol in the bottom right-hand corner of the card. These cards focus on similar data, such as mode. The students discuss the meaning of their card and what information it provides their group.
> The students return to their original group, read and process the data. Students record the reconstructed data and check clues for accuracy.
> The whole class comes together to share the process they undertook in their groups to reconstruct the data. The teacher asks students to reflect on what they have learned through doing this task.
> Students return to their mathematics journal and individually redefine their understanding of the key terms. The teacher asks the students to share how their current definition compares with their previous held understandings.

Together these two tasks allow students to make decisions about the meaning of these various measures, to compare and contrast similar terms, and to consider the properties and meaning of these various measures of central tendency.

Post-sequence Assessment and Evaluation

At the conclusion of the lesson sequence, teachers administered a survey (developed in collaboration with the researchers) asking students to nominate the tasks they had enjoyed most and those they had learned most from, and to explain their choices. The teachers also prepared an assessment task. Some of the items from the assessment are presented below to illustrate the nature of the learning the teachers had anticipated.

1 Draw what this information would look like as a pie graph.

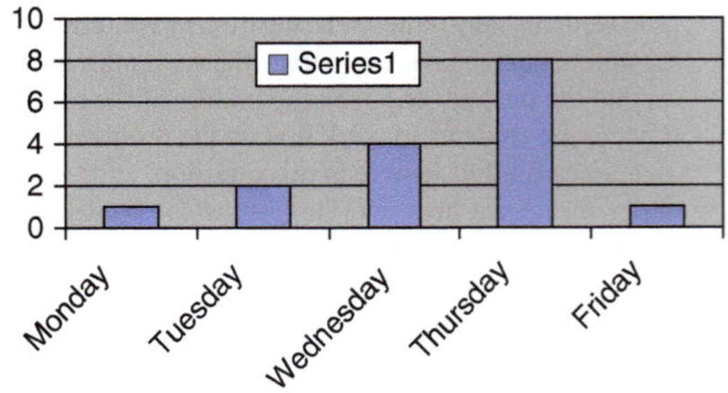

(continued)

(continued)

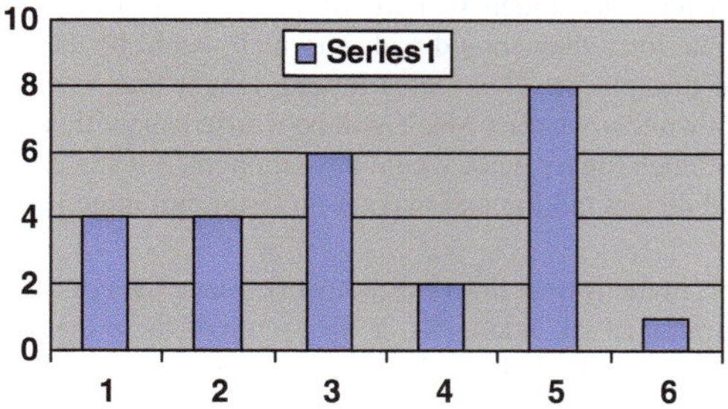

2 What might this be the graph of? Give two different answers, and for each one, put all of the appropriate information onto the graph?

3 Show the following information in a two-way table. How many students are there in this class?

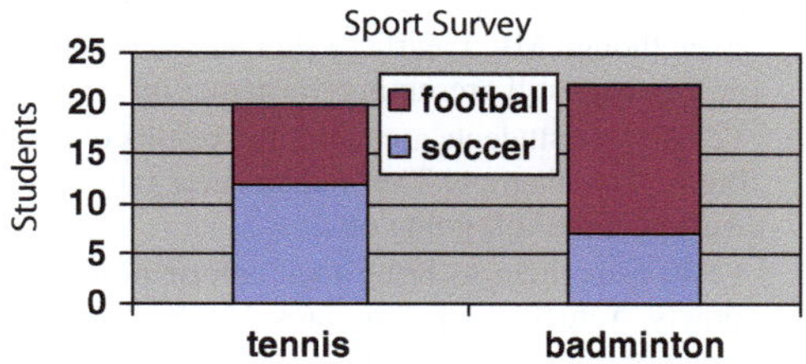

There is some discussion of these in Chap. 10.

The Lesson Observations, Including Observations of Lesson Five

There were detailed observations of each lesson in the sequence that involved structured but open-response notes being recorded in writing as the lesson progressed. These observations facilitated later discussion with the teacher, between the observers, and with the rest of the researchers. The observations provided additional insights into the ways that the tasks operated, the opportunities that they allowed, and the constraints that they presented. The following notes that refer to observations of lesson five are not edited so as to illustrate the nature of the data that were collected in this way. The notes indicate the times that the various events occurred, direct observations, and some interpretations:

> 11:16 having settled them down she got started. With a 2 way table, she posed the questions do you watch the Simpsons, and do you like carrots. She gave them a sticky on which to write their names, they then put them on the two way table. She discussed each aspect of the table and the representations, attempting to ensure that what she is doing is clear for everyone. She set the task that they had to ask two questions, each of which had two answers.
>
> 11:38 they go back to their tables to start work on the questions. Ms. Y had asked them to work individually. It seems the same feature as before in that the students did not try to get started immediately but spent time doing what seems to be irrelevant things. The students near me did not follow the two way table model they were presented with. Some sought to collect information as 4 separate bits.
>
> 11:45 Ms. Y interrupted them to clarify that names were not needed. Some students are using the 2 way tables (which for now I think is the only way to do this) but many are

collecting other information. Many are presenting their table to the respondent, and rather than asking the questions, merely allowing the other student to mark the table. Theoretically, once a student starts using an inappropriate form they should realise (which would be the point of allowing them to do this in the first place). This seems to take longer than one would think, and certainly is more noisy. I wonder whether Ms. Y will be worried about this noise. Ms. Y seems to allow this to go on much longer that I would have thought, but she is talking with students all the time (I would do less talking and make it go faster, but mine is certainly not obviously better)
11:58 Ms. Y shows C's recording which is in the format that does not work (since the cross over is not evident). C is having to go around to ask again. Ms. Y also reminds them that once they have collected the data they have to examine the table for their information.
12:26 after a period of apparent inaction, most of the class (at least 13) are now working on presenting their work, including saying 5 pieces of information they can see in their data. Even though Ms. Y earlier said that she was not interested in names, a number of students near me are still presenting the names on their charts
12:38 some students are recording results on the board, and some other students are drawing a Venn diagram. The students near me—who earlier had it incorrect—have done good work, but are still going on this
12:40 asks them to bring their information to the floor. They straggle to the floor quite slowly. Actually took 3 minutes. Ms. Y asked excellent questions, like: Can we tell whether green or pink is more popular? Can we tell whether soccer or netball is more popular? How many students in the class
She presents the different ways of presenting of information. She introduces venn diagrams—which it looks like they have done previously—She asks How can we transfer this information to the Venn diagram? Again her questioning is excellent and she draws out the mathematics. She has kept them going past the finishing time, and they are restless and but still attending to what is happened.

Some of the aspects which were discussed with the teacher after the observation were the use of time, and especially the pacing of the lesson, the way that she interacted with individuals, the intentions in the post-lesson review, and the extent to which her lesson goals guided her actions through the lesson. The observations are not analysed in detail in this book, but this example illustrates the contribution that the observations overall made to the project.

Summary

While the overall research project was about tasks, teachers created sequences of tasks that variously have the purpose of introducing the content, exploring meanings and connections, and extending the thinking of the students. This chapter presented a lesson sequence that was developed to address a specific aspect of content, using all three of the task types along with an overall thematic task. It illustrates the ways that tasks require students to make decisions, to plan their own work, and to connect different ideas and representations. This sequence also illustrates the advantages of teachers planning and reviewing their teaching together. It also provides evidence that teachers can develop and teach highly effective sequences of lessons incorporating a diversity of rich and challenging tasks.

Chapter 9
Students' Preferences for Different Types of Mathematics Tasks

The focus in this chapter is on the ways that students respond to the tasks as set up by the teacher, which in turn are related to the students' learning. In our project, we sought insights into students' reaction to different types of tasks. This chapter gives a rationale for seeking students' opinions and describes data collection on their confidence and satisfaction, and their ranking of preferences for particular types of tasks. The basic findings are that there is variety in student preferences and opinions and that students distinguish between tasks they enjoy and those from which they learn. It is proposed that teachers need to accommodate this variety in their choice of learning opportunities for students.

Seeking Students' Opinions About Tasks

There have been substantial efforts to seek students' views about aspects of mathematics learning. These include nuanced approaches to students' attitudes (Hannula, 2004; McLeod & Adams, 1989) addressing psychological considerations such as identity, autonomy, and social connectedness, as well as liking, enjoying, and seeing the purpose and potential in mathematics. There has also been sustained study of students' beliefs about the nature of mathematics and mathematics learning (e.g. Leder, Pehkonen, & Tőrner, 2002; Pajares, 1992); the values they attribute to mathematics, the way it is learned, and its uses (e.g. Bishop, 2001); the ways in which students are motivated (e.g. Middleton, 1995; Middleton & Jansen, 2011); and the ways that students connect learning opportunities with how they see themselves (such as whether they can get brighter through effort), and the subject (such as whether effort leads to success) (Dweck, 2000).

We agree with Allen (2003) who argued that there has been too little attention to students' perspectives of aspects of teaching and class organisation despite major changes being made in mathematics curriculum, assessment processes, and pedagogy. Allen conducted semi-structured interviews with a range of students in middle

P. Sullivan et al., *Teaching with Tasks for Effective Mathematics Learning*, Mathematics Teacher Education 9, DOI 10.1007/978-1-4614-4681-1_9,

years' classes, and used both qualitative and quantitative methods for analysing the data. The students she interviewed reported that the set (meaning the streamed or tracked level) into which the students were placed had a significant effect, both positive and negative, on the students' views of themselves as mathematics learners, and that the nature of assessment was also considered important by them, as were the type of rewards that they received. Allen concluded that the students evaluated their mathematics learning in terms of correct answers. This last element is connected directly to the ways students interpret tasks.

In seeking students' views about tasks, we chose to focus our data collection on the extent to which they felt they learned, and whether they liked particular types of tasks, since these seemed to be main determinants of their decisions on engagement. In the piloting of our instruments, we found that the students were able to respond to both types of prompts without requiring further clarification. Our approach was to seek some responses to predetermined scales as well as some free format narratives by the students to allow their real concerns to emerge. This chapter presents two sets of data in the following sections: the first comprises students' responses to particular prompts in a survey (termed the Learning Mathematics Student Survey) and the second includes student ratings and responses to tasks after completing a sequence of lessons planned around the types of tasks we described in Chaps. 4–6.

Responses of Students to Pre-determined Prompts About Tasks and Pedagogies

The survey was designed to gather responses on aspects of lessons and tasks from a cross-section of students. We sought responses from all students from grades 5 through 8 in the project schools, not just those in the classes of participating teachers. As well as seeking information on various aspects of lessons, we included specific items asking students to compare different types of tasks and to indicate their preferences.

The items on general aspects of pedagogy were adapted from Clarke et al. (2002) and Sullivan, Prain, et al. (2009) and the items on tasks were written specifically for this purpose. The survey was piloted with some classes of students in schools similar to those in the project, and we interviewed students in those classes to seek clarification of confusing responses. For example, one of the items asked students to indicate how frequently they used calculators in their mathematics class. In one of the pilot classes, there was a range of responses including some students who claimed never to use calculators, and others who reported they used them in every lesson. We queried these students on how this could be possible and one student replied, "Well I never use a calculator but they (pointing to two other students) use them all the time". After some revision, we administered the survey.

We asked each school to nominate one of the project teachers to co-ordinate the administration of the survey across all classes of students in the target years to ensure that the students completed the survey individually and seriously. The results were entered professionally, including double-checking of the entries. The analysis of the free-format items is described in the next section.

There were 930 students in 96 classes across 17 schools who completed the survey. Note that there were fewer responding students proportionally in grades 7 and 8. This was a characteristic of the project overall, with more teachers at grades 5 and 6 participating, and also representative of the difficulties in seeking survey responses in secondary schools. One particular challenge is that we only invited students to complete the survey who had returned a signed permission form from their parents, and some secondary level students seemed reluctant to do this.

Students' Attitudes

The first two items in the survey asked the students to indicate on a scale of 1–7 "How good are you at maths?" and "How happy are you in maths class?" In the following discussion, we take responses to the first to be a measure of confidence, and the second to be a measure of satisfaction. These items were intended to allow differentiation of later analyses based on the students' responses.

The students' responses were predominantly positive, although they did vary from 1 to 7 indicating that there is a range of levels of confidence and satisfaction with their mathematics. We were surprised to find that there was not much difference between the responses of the students at the respective year levels, given that the junior secondary students appear to an observer to be less satisfied and less confident in their ability. Table 9.1 presents a comparison of responses of students across these year levels.

Overall the students seemed to feel somewhat more confident than satisfied and there is a substantial spread of scores at each of the grade levels, meaning that while there were some students who gave positive responses, there were others who gave negative ones. While there were statistically significant differences between the grade levels for both confidence (F (3,926) = 3.34, $p < .02$) and satisfaction (F (3,926) = 4.11, $p < .01$), the differences within each year levels were more substantial than those between year levels. Rather than teaching differently in grade 5 than grade 8, the implication is that teachers need to address differences in confidence and satisfaction in the classes they are teaching, whatever the level.

We were interested in exploring the relationship between confidence and satisfaction. To do this, we recorded the confidence and satisfaction responses into three roughly equal categories in size. For both variables, scores of 0, 1, 2, 3, and 4 were the bottom third, scores of 5 were the middle group, and the students who indicated 6 or 7 made up the top group. We found that, while the correlation between confidence and satisfaction is strong ($r = .53$, $n = 934$), fewer than 50% of those in the top third of confidence were also in the top third of satisfaction. In other words, it is not appropriate for teachers to make an assumption that because a student is confident they can do mathematics, they will also be satisfied in mathematics classes, or vice versa.

Table 9.1 Comparison of students' "confidence" and "satisfaction" across the year levels

		Confidence		Satisfaction	
Grade level	*n*	Mean	s.d.	Mean	s.d.
Grade 5	302	4.70	1.38	4.40	1.65
Grade 6	392	4.95	1.24	4.28	1.52
Grade 7	126	4.61	1.33	3.82	1.55
Grade 8	110	4.91	1.25	4.15	1.76
All	930	4.81	1.31	4.24	1.60

There were also significantly higher levels in boys' responses, compared to girls', for both confidence (F (3,926)=4.11, $p<.01$) and satisfaction (F (3, 926)=3.34, $p<.02$), but again the within-group differences were more educationally important than the between-group differences. In other words, the diversity of confidence and satisfaction within the girls and within boys is a more urgent issue for teachers to address than the differences between boys and girls.

A more startling result is that there was a range of means for the classes of different teachers. Using only classes with more than ten responses, the range of means for confidence was from 3.9 to 5.6, and for satisfaction from 3.7 to 5.2. These are substantial differences, and it is suspected that teachers have both positive and negative effects on both student satisfaction and confidence with mathematics. It would be useful for teachers to try to establish students' levels of confidence and satisfaction and to seek to address negative responses by specific actions.

We also sought students' responses to different types of tasks. There were two sets of tasks presented, and for each set, students were asked to rank the tasks in terms of which they liked the most, second most, etc., and which they considered they would learn from most, second most, etc. The specific instructions were in the following form:

> In this table there are three maths questions that are pretty much the same type of mathematics content asked in different ways.
> *We don't want you to work out the answers.*
> Put a 1 next to the type of question you like to do most, 2 next to the one you like next best and 3 next to the type of question you like least:

The tasks were then presented. The labels in parentheses used here were not presented to the students, but are shown to assist in interpreting the tables of student responses. Note that these tasks do not match our types strictly. The first two were just routine examples, whereas we intend that our tasks do more. For the survey, though, we were seeking to have the information familiar to the students' experience. Nevertheless, the three examples do reflect the respective emphases and style of our tasks. The questions were not presented in this sequence.

Number calculation (representational)	$2\times13+4\times7=$
Word problem (contextualised)	Movies tickets are $13 for adults and $7 for children. How much does it cost for 2 adults and 4 children to go to the movies?
Open-ended (open-ended)	2 Adults and 4 children went to the movies. They spent $120 on tickets. How much might the adult and children's tickets cost?

Table 9.2 Students' first preferences for "like" and "learn" (%) ($n=922$)

Task	Like most	Learn most
Number calculation	54	40
Word problem	35	23
Open-ended	12	37

The students were asked first to rank the questions according to how much they would like to do them. They were then asked to rank them according to how much they felt they would learn from them. Note that we consider the first two to be straightforward examples, and we assume that the third question would appear to be more difficult for the students. Table 9.2 presents the percentage of students who chose the respective tasks for their first preference for each of "like" and "learn".

Each of the tasks was liked by some students, although the most often liked task was the number calculation. In order to explore the reasons for their ranking, the students were asked to respond in their own words to the prompt:

I like to do this type of question (the one you put a 1 against) the most because

Their written responses were typed, inspected, and categories determined; the responses were then coded; and the categories refined. There was a range of categories for their responses, but we only present here examples from the particular category that represents the majority of the responses in each case.

For those who chose the number calculation as the one they like, 72% of the 497 responses were coded as "because it was easy". Note that these were their own words, and there were many possible categories. Specific examples of what they wrote are as follows:

Because it's quick and easy for me to do.
I'm confident with that type of maths and I know more about it than the others.
I'm really good at times tables so I can get this answer the quickest and its easier.

For those who chose the word problem, 58% of the 317 responses were also categorised as "because it was easy", giving similar types of responses.

For those who chose the open-ended task, the most frequent category was "challenge" which was chosen by 33% of the 113 students, giving reasons such as

I like to do it because the other ones don't really teach you much
They are challenging and I like challenges
I like it because I want something challenging and hard

It seems that the students who liked a task because it is easy (which was over half of the students) were more likely to choose the number calculation or the straightforward word problem, while students whose reason for liking was that it was challenging were more likely to choose the open-ended task.

We were also interested in whether particular types of students were more likely to like particular tasks. Table 9.3 compares the responses of the third of students who rated themselves as lowest on the confidence scale, and the third who rated themselves as highest.

Table 9.3 Comparison of responses of students' confidence and the task they most liked.

Task	Lowest confidence	Highest confidence
Number calculation	44	60
Word problem	42	26
Open-ended	12	16

Contrary to our expectation, the students who rated themselves in the top third of confidence were more likely to choose the number calculation task. Perhaps they feel that they are good at such tasks, or perhaps being good at such tasks is why they rated themselves as confident. Also of interest is that the students who rated themselves as the least confident were more likely to choose the word problem as the one they liked. While there is some evidence that suggests that contextual problems are more difficult for less confident students (e.g. Bransford, Brown, & Cocking, 1999), it also makes sense that such students might prefer a more obvious rationale for doing mathematics, and word problems provide that.

The second aspect of the data presented in Table 9.2 relates to the tasks from which students felt that they would be most likely to learn. The number calculation and open-ended tasks were rated as equally likely to help them learn, more so than the word problem, and again, each of the tasks was chosen by some students. We were also interested in the reasons that they gave for their choice. The prompt was as follows:

The reason this type of question (the one you put a 1 against) helps me learn the most is:

As before, the responses were typed, inspected, categories developed, the sentences and phrases coded, and then the categories revised.

Forty-four per cent of the 369 students who rated the number calculation as the one from which they were most likely to learn gave responses that were coded as challenge, writing statements like:

It helps me with my addition AND multiplication.
I am not good division, it teach me more.

Even though there was a trend towards choosing this type of task because it appears to be easy, it seems that there are also many students who feel that this is what they need to learn.

Nevertheless, there were also 28% of these 369 students who claimed to learn most from the number calculation "because it was easy", writing comments such as:

I know what to do and I do it all the time
It gets straight to the point
It's not as confusing as the others

Perhaps these students are like those of Allen (2003) who connect learning with correct answers

Of the 213 who chose the word problem as the one from which they would learn most, 42% were coded as "challenging", writing statements such as:

Because it's difficult you have to read it good
It is more hard and challenges me more

Most strikingly, of the 342 students who chose the open-ended tasks, 79% were coded as "challenging", writing statements such as:

> Because it's tricky and if you do you can do a lot more
> It makes you explain it more
> It's difficult so I think more

The top third of the "good" students were more likely to choose the open-ended task as the one from which they would learn.

Interestingly, few students chose a particular task as first preference for both liking and learning, indicating that most students see liking and learning as different (as they did for confidence and satisfaction).

The strongest association here is that well over half of the students overall, irrespective of the task type they chose, connect learning and being challenged. We emphasise that the students were not responding to directed prompts and were merely being asked to give reasons for seeing a learning opportunity in a task, and we argue that this provides a meaningful insight into their opinions.

We also asked the students to rank a second set of tasks. The students were given the instruction that each of the tasks involved finding an area. The students were asked to rank the tasks similarly, first for "like" and then for "learn". In this case, the more difficult task is the practical calculation:

Area by counting (representational)	Find the area inside the following shape (with an irregular shape overlaid on a square grid presented in a diagram)
Practical calculation (contextualised)	A running track has straights that are 100 m long with half circles at the end of the grass. What is the area of the track? (diagram of a running track presented)
Open-ended	A shape has an area of 10 square units. What might the shape look like?

Table 9.4 presents the percentage of students who chose the respective area tasks for their first preference for each of "like" and "learn".

In this case, more students liked the representational and the open-ended task. Note that few of these students would have learned to calculate the circumference of a circle and so the practical calculation would have seemed very difficult. We note again, though, the variety of preferences: each task was nominated by some students as the one they liked most, and similarly each task was nominated by some students as the one from which they felt they could learn most.

We also asked students to state their reasons for these choices, and the types of comments they made were similar to the ones they made for the previous set of tasks. Again there were some strong trends in their responses, given that these were not prompted, and there was a range of possible ways that the responses could be coded.

There were 49% of the 358 students who, having selected the *area by counting* task as the one they most liked, gave the reason "because it was easy". Of the 179 who most liked the practical calculation, 39% wrote "because it was easy", and 30% said it was "challenging". Of the 382 who most liked the open-ended task, 43% said this was because it was easy. The third of the students rated as most confident were more likely than the low group to choose the open-ended task, and both the low

Table 9.4 Students' first preferences for area tasks for "like" and "learn" (%) (n=922)

Task	Like most	Learn most
Area by counting	39	33
Practical calculation	20	44
Open-ended	42	22

group of confident and the top group of satisfaction were less likely to choose the *area by counting* task.

Table 9.4 also presents the results for the preference for the task that would help them learn. In this case, most students reported they would learn most from the practical calculation.

Of the students who rated the respective tasks as the one from which they learn best, 54% of the 305 who chose *area as counting* said it was because it was "challenging", as did 75% of the 408 who chose the practical calculation, along with 33% of those who chose the open-ended task as the one from which they would learn. The "good" third of students were more likely to choose the practical calculation.

Recognising that it is hard to see how the students could rate the practical calculation as easy, while it may be possible for some students to see the *area as counting* and the open-ended one as easy, these respective tasks may also appear difficult to others, given that there is a less certain answer. Again it seems that while there is a diversity in the tasks the students like and learn from, a majority of students indicated a liking for tasks that they feel are easy, but gave a strong indication that they consider themselves more likely to learn from tasks they feel are challenging.

In Summary

To summarise the results from these items on the survey, it seems that at each of these middle-years' levels, there is a range of student satisfaction and confidence, and teachers should be aware of the views of each of their students. It also seems that teachers make a difference to students' responses and that teachers need support not only to find out students' levels of satisfaction and confidence, but also on strategies to address negative responses. Each of the task types is liked most by some students, and likewise each of the types is rated as the one from which they can most learn, and this suggests that teachers need to use all types of task in their teaching. This may be particularly relevant to teachers at the secondary schools who seem to use texts with a limited range of task types very frequently. A related issue is that students may need support to gain benefits from tasks that they do not like or do not feel that they can learn from. It seems important that teachers make students aware of the purpose of tasks and what it is that the teachers are hoping the students will learn from them. The students seem to like tasks that are easy, yet feel they learn best from tasks that are challenging. Of course, we would hope that students can also learn from tasks they find easy, and like tasks that are challenging. Again it may be important for teachers to illustrate or emphasise the role of the tasks and the nature of the challenge they offer.

Tasks Preferences Within a Lesson Sequence

As described in Chap. 8, we developed sequences of lessons with teachers in the style described by Ruthven, Laborde, Leach, and Tiberghien (2009). In each case, the teachers, after volunteering to participate in this aspect of the project, nominated a topic and a period of time. The teachers then met with the researchers to discuss ideas and possible activities. We asked the teachers to consider a spread of the task types, but the decisions on the structure of the sequences were theirs. Each group of teachers developed detailed plans which outlined their intentions for the sequence. The following presents some of the data collected in one of these sequences, focusing on the students' responses to the tasks.

This sequence consisted of 15 lessons, each of which was planned in considerable detail, and each of which was timed to last around 90 min. The focus of the sequence was statistics.

At the end of the sequence, we invited students to indicate on a list of the 15 main tasks they may have completed which one they liked the most, and the second most, and from which one they had learned most, and the second most. Table 9.5 presents the name of the task, and the number of students, out of the total of 50 responses, who indicated these preferences respectively. The tasks were those presented in Chap. 8.

The most striking observation from these data is the diversity of responses, in liking across tasks and in learning across tasks. The spread of responses was across tasks and across task types. There was no one task that had majority support. There was a similar diversity of students' preferences in the other two lessons sequences as well. No one task failed to appear in anyone's first or second preference. Perhaps a major result from this project is in identifying the diversity of student preferences for tasks both in their liking and perception of learning. The explicit implication is that teachers need to consciously choose a variety of types of tasks when planning learning sequences.

There was some ambiguity in what students took to be "learning". For example, while *Rock paper scissors* was liked most, it was low in terms of whether students considered they had learned from it. Neither of the two students who did choose "learn" for this task had indicated they liked it. While it was a familiar game, it was the only task that no one chose as their first preference for learning. Yet to complete the task, students had to make substantial and complex decisions on their method of recording results, and they were required to report on the outcomes of the 100 trials. There were unanticipated challenges in that most students in recording their results did not consider how they would then represent them graphically. In other words, it is difficult to imagine that students did not gain insights into recording of results as a result of the task. As with other aspects of these student data, we suspect that the meaning that students give to the term "learn" is subtle, and requires further exploration.

We were, of course, interested in whether there is connection between their task preferences and their learning. To do this, we analysed those assessment items that reflected a particular classroom task. The following analysis is for just one particular

Table 9.5 Student preferences for "like" and "learn" for the Chance and Data activities

	"Like" 1st preference	"Like" 2nd preference	"Learn" 1st preference	"Learn" 2nd preference
Finding similar graphs	0	0	1	3
Clues on cards	0	3	9	5
Excel to present data	7	3	6	3
Rock paper scissors	11	14	0	2
Matching graphs	1	0	1	5
2-Way tables	1	4	3	3
Average height	0	3	4	1
This goes with this	2	1	2	7
Most common letters	3	1	4	1
Av ht in school	13	9	4	8
Sentence with 5 words	2	6	4	2
7 People went fishing	5	2	9	6
Conducting a survey	1	1	2	1
Weather project	4	3	1	1

assessment item and task, but the results were similar on the other items. The item on the assessment was as follows:

> Five teachers at [the school] have a netball goal shooting competition, with each of them having 10 goals. So each teacher ended up with a score out of 10. The mean number of goals scored was 6, the mode was 4 and the median was 5. Write down what the five teachers' scores might have been to make this work. (give at least two different answers)

If the students gave one or more correct responses to all three statistics (mean, median, and mode), they scored two marks; if they were correct on two of three statistics, they scored one mark; otherwise they scored 0.

The equivalent task in the lesson sequence was as follows:

> Seven people went fishing. The mean number of fish they caught was 5, the median was 4, the mode was 3. How many fish might each of them have caught?

To examine the extent of any relationship between the students' responses on the assessment and their responses on the task preference survey, Table 9.6 presents the number of students who scored the respective assessment scores (that is 0, 1, or 2) and the number who rated the classroom task as the preference for the task they liked first, or second most or did not rate it at all. The numbers in brackets refer to the number of students who rated this task as the first preference for having learned from, second preference, or did not rate it.

Note that this task is quite complex for students at this level, both for the class task and for the assessment, and the level corresponds roughly to the curriculum 2 years later. That 27 of the 47 students were able to give reasonable responses to the assessment item, given its difficulty and the strict scoring rubric, indicates that many students did learn about mean, median, and mode from these experiences. Of course, it is possible that students did learn but still could not complete the assessment task, and other students did not rate the task in the top two for learning yet

Table 9.6 Number of students in project teacher classes who rated the task in terms of liking (learning) based on assessment scores for the fish task

	0	1	2
1st preference	3 (5)	1 (2)	2 (3)
2nd preference	1 (3)	0 (0)	1 (3)
Did not rate it	16 (12)	5 (4)	18 (15)
Total	20	6	21

scored 2 (perhaps they knew it already). It is suspected though that students see liking and learning as different, and perhaps that if they like something they believe that they cannot be learning (and maybe even vice versa).

There are no clear relationships in the data presented in the table. Three students who claimed to most like the task did not score on the summative assessment. The five students who claimed they most learned from this task also did not score on the assessment item. Yet there were 18 students who completed the assessment task fully who liked other tasks better (15 such students for "learn"). The students' workbooks were also examined and a score allocated based on the completeness of their in-class written response to this task, ranging from 3 (complete) to 0. A similar analysis to the above was undertaken and no trends were apparent there either. Clearly this analysis has not revealed insights into any connections between students' stated preferences for liking and learning from tasks and their achievement.

In summary, the analysis of students' responses to tasks from the lesson sequences emphasised the diversity of students' perceptions of tasks that they like and feel they can learn from, and it is essential that teachers not only use a variety of task types but also explain the purpose for each of the types they do use. There was also no connection between students' stated preferences and their test scores, or indeed the record from their workbook. It seems that what students mean by "like" and "learn" is subtle, and may be only connected loosely to what they do learn. This requires further investigation.

Summary

This chapter has reported data that provide the perspective of students' on the types of task that we have described in Chaps. 4–6 of this book. The data indicate that in all classes, there is a range of levels of student confidence and satisfaction, and that the key variability is between students rather than between classes. A common finding across the different data that were collected is that there was a diversity of students' responses to the types of tasks that students claim to enjoy and from which they report they can learn. While the reasons for the differences between their enjoyment and learning are unclear, an obvious implication is that teachers need not only to be explicit about the purpose of each task that is posed, but they also need to pose a variety of types of tasks to cater for the diversity in interest.

Chapter 10
Students' Perceptions of Characteristics of Desired Mathematics Lessons

The third stage in the Stein, Grover, and Henningsen (1996) framework is mathematical task as implemented by the students. In addition to seeking students' opinions about the types of tasks they like and can learn from, we also sought their opinions on the types of mathematics lessons they prefer. It was hoped to gain insights into the ways students described their desired characteristics of lessons, rather than through their ratings of lesson characteristics prepared by us. We did this through open-ended responses to particular prompts on the overall survey and free format essays by students in two of our project schools. We based this approach on Zan and di Martino (2010) who argued that researchers should move from measuring attitudes to describing them. They argued for more narrative approaches to describing student attitudes, including with large samples, with the goal of understanding behaviour. This chapter seeks to elaborate factors which it can be argued might influence student behaviours. Three key recommendations relate to the forms of student grouping, the explicitness of teachers' intentions, and the ways that teachers interact with students.

Responses from the Overall Survey on Desired Lesson Characteristics

In addition to the data reported in the previous chapter, the student survey sought some free format responses on items. These were administered in a similar way to the items discussed in Chap. 9. One of the prompts on the survey asked the students to write a free format response to the following:

> Think about all the maths lessons you have EVER BEEN IN. Now think about the best maths lesson you have EVER BEEN IN. Describe what you did in that lesson.

P. Sullivan et al., *Teaching with Tasks for Effective Mathematics Learning*, Mathematics Teacher Education 9, DOI 10.1007/978-1-4614-4681-1_10,

Table 10.1 Number of student comments in various categories of the "best" mathematics lesson (n=930)

Category of response	Total mentions
Game that taught us maths	184
Competition or test on maths we know	83
Outside activity	59
Particular topic, e.g. Measurement	395
Real-life problem, e.g. Water in tank, maths to make food	49
Used or made a model, e.g. Pita bread for fractions	258
Particular operation, e.g. Multiplication	119
Learned mathematics I didn't know before	16

The responses were generally informative but brief. The following are examples of the student's responses.

> We played the grand final of maths football. You have to answer questions and if you get them … the football goes closer to your goal. We won.
> We did long jump outside and then we measured everyone's jump and then put it in a chart.

The responses were read and preliminary categories for grouping the students' responses identified. The various responses were then coded by a second researcher, and adjustments made to the categories. To indicate the types of responses given by students, and the ways that we applied the codes, the following are some illustrative sentences and phrases allocated to particular categories. We coded as

Game that taught us maths general statements such as "we played maths games on the computer", and more specific statements such as "we coloured in some boxes on a fraction we roll two dice whatever fraction you get you colour in".

Particular topic statements such as "we added fractions. We learnt how to add them with different dimoniaters (sic)", "algebra would be the best lesson because I was good at it", "I liked percentage. At the start of the term I couldn't understand but when my friend's and my teacher helped me it became easy as + and –", "when I was learning about decimals", "learnt how to add and subtract mixed numbers and turn them into improper fractions".

Particular operation statements such as "I was starting to learn multiplication and I got it so easy and I loved it".

Used or made a model general statements such as "when we did hands on activities" and more specific comments such as "smarties maths. We used smarties to work out fractions (colours). It was really fun!", "when we were making the maps of a town with 24 houses", and even (!) "when we drew Cardoids, Mystic Roses Hyperbola".

The number of comments coded in the various categories are summarised in Table 10.1. Note in more than one category.

Table 10.2 Combined categories of responses to characteristics of best lesson ($n=930$)

Category of response	Total mentions
Those that focus on pedagogies	633
Those that focus on content	530

Table 10.3 Number of students in combined categories broken down by self-rating of confidence

Category of response	Low third ($n=337$)	High third ($n=292$)
Those that focus on interesting pedagogies	233	187
Those that focus on content	174	177

There is clearly a spread in the lesson elements that the students chose to mention. The most striking aspect is the diversity of types of responses. We had perhaps anticipated that students would like interesting aspects of pedagogy such as games, real-life problems, and use of models, but were surprised at the number of responses that focused on a particular topic. To explore this further, Table 10.2 presents the above categories combined into those that could be interpreted as focussing on content, and those that focus on pedagogy.

In other words, nearly half of the students in describing their "best" mathematics lesson referred to specific content, and just over half mentioned interesting pedagogies. This surprised us. To explore whether a reference to a topic was a characteristic of a particular type of student, Table 10.3 presents the number of responses in each of the categories in the previous table given by the students who rated themselves in the lowest third of confidence, and those who rated themselves in the highest third of confidence.

There is a slight tendency for the students who rated themselves as more confident to mention a topic. Nevertheless the interesting feature is that many students who rate themselves as not confident at mathematics mention a topic, while around 40% of the students who rate themselves as more confident referred to a game, real life, or a model.

These tables, taken together, indicate that crafting lessons out of tasks is not only through creative pedagogies, but also can be through a focus on content. We suspect that finding interesting ways to help students learn a particular topic is the ideal combination. Indeed that is exactly what the tasks described in Chaps. 4–6 are intended to do.

To explore the reasons behind the students' descriptions of their best lesson, we also invited them to answer, in free format:

> Why did you choose that as the best maths lesson ever (that is, what made it good)?

Again categories were created and progressively refined. The following are the categories that seemed to capture the major themes, along with an illustrative example of some students' statements:

Challenging: "It was one of the most challenging maths lessons, and the feeling of achieving the answer was great", and "I like that maths lesson because you had to think".

Table 10.4 Frequencies of particular responses explaining their choice of a best lesson ($n=930$)

Category of response	Number of mentions
Challenging	89
Easy	65
Fun/interesting	502
Learned something new	179
I'm good at this	101
Went outside	47
Worked in groups	51
Made a model	80

Easy: "I chose that one because it wasn't too hard for me", and "Because we didn't have to do any work".
Fun/interesting: "Because we had fun", "It was entertaining and fun because it was a race to win a game".
I learned something new: "Because I learnt how to times decimals".
I'm good at this: "Because I was the only person that knew it", "I chose that as the best maths lesson because I'm really good at".
Went outside: "We got outside, in the fresh air and did real life mathematics", "We got to run around and use our brains at the same time".
Worked in groups: "I really love working in groups and I think that working in groups makes me think better", "We got to do stuff with friends".
Made a model: "We did more 'hands on' than paper and pen", "I love creating things and we got to make a robot".

Table 10.4 presents the frequencies of statements coded in this way for this prompt:

Again the diversity of comments is noteworthy. The most frequently used code, with around 45% of responses overall, was the fun category indicating that this is indeed an important aspect of planning mathematics learning experiences in the eyes of the students. There were 369 responses that referred to learning something new, to facility at the task, and to challenge, indicating that some students respond to these aspects as well. Again, it seems that the advice to teachers is to devise "fun" ways for students to learn something new while they are being challenged.

In summary, the main impression from these responses is their diversity, and there are clearly many ways in which students respond to lessons. There were two trends in their lesson descriptions of, on one hand, students recalling effective teaching of a content topic, whereas there were others who remembered interesting aspects of the pedagogy. In explaining their choice of lesson, the main category of responses related to fun, but learning something new was also frequently cited. Teachers are therefore advised to focus on the students' learning of content, and to choose interesting and fun ways to engage students in that learning.

Students' Essays on Their "Ideal Maths Class"

At a different time, and in a different way, we also sought students' views on lessons and teaching through a particular prompt seeking narrative responses. We asked students of some of our project teachers to write an essay, the particular prompt of which was as follows:

> Write a story about your ideal maths class. Write about the sorts of questions or problems you like to answer, what you like to be doing and what you like the teacher to be doing in your ideal maths class.

The intention was to gain insight into what the students recalled about their mathematics classes, and it can be assumed that these responses can be taken as indicative of the lesson features that the students like. The following are three examples of typical students' essays, presented exactly as they were written:

> My favorite maths would start with a 10 min introduction where the teacher explains the game to all of us and still allowing time for questions. The games would be 2+ people for a competition and people will split into groups and will organize who plays who 5 min every one will be playing at all times unless there is an odd amount of people we will play for 25 min. at the end of the lesson the groups will figure out who was the winner and people can share strategys they used. Sharing is for 10 min. For my second option I would do real life problems Like 250 grams of sugar for $10.50 or 750 grams for $33.15. I like real life problems because they could help me one day and its set out differently than math. For this the explanation is for 5 min this is because you don't need to explain the rules.
>
> My ideal maths class would be hands on work where we would make something. But we would use lots of different types of maths for example measurement, sums, fractions and percentages. It should be challenging and we would work in groups. If we got stuck on something, we would sit down with the teacher and work it out together. The thing that we make should be something we could use in our every day lives. It should also need to be researched on the computer and maybe in books. We would not use calculators because I don't really learn anything from them.
>
> When I do maths I enjoy doing hands on activities and outside activities. I enjoy hands on activities because sometimes I get to eat it like fairy bread fraction, chocolate fraction and I enjoy outside activities because we need to go outside once in a while and I enjoy maths that has lots of answers to it and I like group work because I can contribute with other people and I like my teacher because she helps us for our maths.

In the first response, there were two key elements: the use of a game and the use of real-life problems. In the second, the focus is on addressing content. In the third, the focus is on pedagogies.

The instructions to the students seemed to be different in the schools, in that students in one of the schools were more likely to give an answer in point form, such as the following example of one student's response:

> Activities like the 4 fours.
> I like challenging activities.
> I like working on the computer sometimes.
> Working in the learning common.
> Activities with a small group or a partner.
> An activity where we have different ways we can present our work.
> I like having a partner and presenting my/our work on a big poster.

Working in small group.
Hands on activities.
Going outside to do activities.

These aspects are somewhat eclectic. He or she likes challenge, working in groups, presenting products on investigations, hands on activities, and going outside.

Each of the responses was read, and a preliminary set of codes established. A second reader then used the first set of codes and added others as appropriate. Where a sentence or phrase could be categorised in two places, this was done.

The first characteristic of the responses is the diversity of aspects on which the individual students commented, again suggesting that there is no commonly agreed ideal lesson, and there are many ways to teach well. The following is an attempt to communicate trends or themes in the responses. Since it seemed that there were aspects of the responses idiosyncratic to particular classes and schools, the responses from the two schools are discussed separately.

In the outer suburban school, there was a total of 39 students in two classes, one Year 5 ($n=21$), the other Year 6 ($n=18$). Unless otherwise stated, the responses were from both classes. In this school, the Year 6 class mainly responded in point form, and so multiple aspects were mentioned by many students.

Thirty students included a desire for materials in their ideal lessons, and some mentioned specific examples such as teddy bears, robots, alarm clocks, class market, and mapping. These are not the structured materials that teachers would naturally expect to see in such responses. There were also 25 specific references to working outside (12 Year 5, 13 Year 6). Given that one suspects that this happens quite rarely, this is a surprising result. Note also that some students in the survey also mentioned going outside as their best lesson. Fourteen students mentioned a connection to practical aspects such as food, money, and newspapers.

There were 25 specific mentions of working in groups as part of their ideal class, and a further 15 mentions of working in pairs. Note that there were also nine students who wrote that they preferred to work alone. The ways of working in class are clearly important for students, and however the teacher intends that the student work, the reason for this needs to be clarified for the students.

There were 22 students who wrote that they liked to be challenged, and 15 students wrote that they liked open-ended tasks or those that had more than one answer, although all 15 were from the Year 5 class. Twenty-one students liked to be helped by the teacher, although most of these were also from Year 5. There were also nine students from this class who mentioned fractions, which is notable in that specific topics were seldom mentioned by any other students, in contrast to the free format survey responses.

Sixteen students wrote that they liked clarity in both the lesson goals and teacher explanations. Examples of such responses were "makes sure we understand", "gives examples", and "explains focus of lesson".

In the suburban school, there were three mixed Years 5 and 6 classes involving 65 students. There were similarities in the responses of the students. Fifty students wrote that their ideal lesson included working in groups, in pairs, or with friends. Twenty-six students claimed to like a challenge, and 22 liked problem solving.

Nineteen specifically mentioned fun or enjoyment, 22 mentioned games, 17 mentioned specific hands-on activities, 13 mentioned specific measurement activities, and 15 gave specific examples that connected learning to their lives. Similar to children in the other school, 18 students saw their ideal lesson as being outside.

Both schools were technology rich, and it was interesting that this was mentioned by very few students. Perhaps they saw the availability of technology as a given, but the scarcity of such mentions may require some further investigations.

In summary, it seems that the responses to this prompt about an ideal lesson seemed dependent on the teacher. In synthesising the responses, students liked lessons that used materials (although these were not structured materials), were connected to their lives, involved games, were practical with some emphasis on measurement, in which they worked outside, and with the method of grouping being important; and over half of the students claim to like to be challenged.

Some Themes in the Data

The following discussion considers some themes in the data using the combined responses from the two schools.

Classroom Grouping

Classroom grouping was an important issue for the 110 children who wrote about their "Ideal Maths Class". While 79 explicitly mentioned grouping, working collaboratively was also implicit in many other examples, such as when students advocated playing games, or measuring one another. Only 17 wanted to work independently, but 12 of these sometimes wanted to work collaboratively. As a grade 5 student wrote: "I like to be on my own or with a partner but on problems I think its better to work in a smaller group". A classmate suggested they "have group, partner and solo work so you can learn in every way and so it fits everyone's needs". Thus there were just five students who wanted to work independently all the time. Typical of this group was a student, who wrote, "My ideal maths class would be problem solving tasks or something to do by myself because I don't like working in groups". These five students were also, in their own estimation, good at maths. The mean of their responses to the general survey question "How good are you at maths?" averaged 6.0 on a scale of 0–7, compared to the 110 essay writers (mean = 5.45) and the mean of all grades 5 and 6 students in the survey (4.83).

One student, with a score of 7 for both "good at maths" and "happy at maths" on the student survey, wanted to be as autonomous as possible in relation to his teacher, his classmates, and his mathematics. This is what another student wrote.

> Introduction: My perfect introduction would be if the teacher who was teaching us on times tables. Would explain the whole session and then send us of to think of ways to practice and then do them. The teacher wouldn't give it to us we think of our own.

> Working Part: My perfect working part is when everybody does their own work and the room is quiet. Because it gives the best opportunity to do your own work and show it to the teacher and she knows it was done by you because you were by yourself.
> Reflection: My favourite reflection in maths is to do a quiz at the end of a session to show how much you've learnt.

This response was unusual, however. The vast majority of these students expressed a desire to work in groups or in pairs and clearly articulated reasons for preferring particular groupings. Overall, 74 of the 110 children expressed the wish to work either in a group or with a partner. Eighteen preferred to work with a partner or friend, while others' preferred group size varied from 3 to 6. Their reasons varied. Some wrote of enjoyment:

> I like working in groups because it is much fun.

Or the challenge of working with others:

> I also like to work in groups because I like a challenge and other people are challenging.

Children often reported that collaborative work was more efficient:

> In all of the lessons people have a group to work with so that it is easy and quick so you come together at the start and find things to do so its quicker.

Some also said that they like to share strategies:

> I like working in big groups or with a partner because we can talk about different ways to solve problems.
> I like being able to chat a little to friends and get ideas for your work. And it makes work time more fun, just don't get carried away with a discussion and forget about your work.

And find help close at hand:

> I like to work in groups or partners because some things are to hard to do by myself and I like to work in groups so we can help each other.
> I feel that when people are in groups they feel more Challenged and if your stuck you could just ask someone in your group to explain the task.

Some students suggested that sharing responses can help them to contribute:

> I think small groups are good because you can have a discussion and hear everyone's views and some people feel more comfortable to contribute.

But not all students want to collaborate in this way:

> Work with people at a similar level in maths to not be dragged down by people who don't understand.

Most students in this cohort wanted to work collaboratively, at least some of the time, and they were clear about the advantages and disadvantages of various groupings for their learning and enjoyment. This is not surprising, for the teachers whose classes completed this writing task had been observed by researchers as they organised their students to work individually, in pairs, or groups, sometimes selected freely and sometimes in ability groups, according to the demands of the lesson. Their writing reveals both explicit and implicit advocacy of teacher–student interaction as well as student interactions. This returns us to the fourth characteristic of effective teachers (see Chap. 3), which is to

> interact with students while they engage in the experiences, encourage students to interact with each other including asking and answering questions, and specifically planning to support students who need it, and challenge those who are ready (interacting and adapting).

The students in this cohort were practised in collaborative learning and shrewd in judging the merits of particular groupings in providing for their individual learning and enjoyment. Tentative learners liked the availability of help; others enjoyed the company or the opportunity to share learning strategies. They also had opinions about the teacher's role. It is clear that decisions on the methods of grouping are important when planning lessons based on the use of interesting tasks.

Interaction Between Teacher and Students

In writing about their Ideal Maths Class, it is unsurprising that over half the cohort ($n = 63$) wrote about their teacher's actions, as they had been prompted to comment on this. The total of 106 mentions of teacher actions focused on explaining the mathematics ($n = 52$), helping students (39), correcting work (9), and listening to and understanding students (5). Most of these students simply recorded that in their ideal lesson the teacher introduced the task, but others detailed the kind of introduction they found helpful.

> I like it when the teacher writes what we are doing on the board while telling us what we are doing so if we get off track we can look at the board to get us back on track.
> I like a short introduction, you know "Go to the table, do these sums!" If it's short I won't get bored.
> I like it when the teacher says what we have to do and does a example first and then lets us go do the work. I also like it when they help you and explain it again and when we figurer one out together.

The 39 students who mentioned that the teacher helped students most ($n = 21$) wanted the teacher to be available to help when needed, or to circulate to see which children needed help ($n = 14$).

Some of the children revealed their need for help or understanding:

> The things I like to be doing are trying my hardest to get done what has to be done and waiting patiently for the teacher to come to me.
> I would like the teacher too: Listen to what we have to say (already does).

Others were more independent, requiring the teacher's help only when the mathematics was difficult, or attributing need for help to other students:

> The teacher will help us with things that we don't know like hard multiplication and algebra.
> My teacher would be on the floor with a group of people if they don't understand how to do the work.

Only one of these students commented on the attributes of the teacher ("if the teacher is too strict it becomes boring").

Summary

The data reported in this chapter are from students' comments about the lessons they prefer. It seems that substantial numbers of students prefer lessons that have engaging pedagogies, but there are also substantial numbers who claim to prefer a focus on the content. Teachers need to find ways to combine both: that is, the need to focus on content using engaging pedagogies.

There are three predominant themes in the students' responses. The method of grouping is important for students and so teachers need to explain the method of grouping they are using to the students. The ways that teachers interact with students is also important for them, and it would be useful for teachers to find out the types of interactions that individual students find helpful. There was also a tension between their liking of an approach and the extent to which they feel it helps them learn. This also suggests that teachers need to be explicit about their intentions about these aspects and also to seek students' views.

A further characteristic of the students' responses was their maturity. For example, the students were asked at the end of the survey:

> Imagine you have been invited to speak to some people who are training to be teachers. What advice would you give them about how they should teach mathematics?

In summary, the students advised prospective teachers to:

- Be nice, friendly, patient, calm, enthusiastic
- Know the mathematics, and encourage student self-confidence and perseverance
- Communicate well; make the learning interesting and give clear explanations
- Listen to students; ensuring they understand, and encourage them
- Make lessons fun, challenging, while giving students choice; autonomy
- Pose tasks of appropriate or increasing difficulty, and teach a variety of tasks/levels, topics/tasks/equipment

In other words, the students are possibly as helpful a source of advice for teachers as are teacher educators.

Chapter 11
Contrasting Types of Tasks: A Story of Three Lessons

At one stage as part of the professional learning programme, we included the presentation of an example of each type of task with the same content focus. This component was intended to provide opportunity for reflection on the respective task types and the teachers were asked to consider and explain the order in which they would teach the tasks. This resulted in considerable discussion and an opportunity for reflection on differences between the task types and their place in a logical teaching sequence. We decided as a research team to pursue this further. This chapter reports a particular data collection exercise focused on contrasting the three types of tasks.

It was agreed that we would provide the teachers with tasks and suggested lesson structures for three lessons, one of each type, focusing on the same content. The teachers could choose the order in which they would teach them. Through the implementation of the three lessons, the choice of order, and the teaching of the lessons over consecutive days, it was hoped to gain greater insights into the pedagogical decisions and different aspects of implementation. In addition, data from the children about their preferences in relation to the three lessons were sought. The design created an opportunity for reflection on the differences between task types in their implementation.

The teachers whose feedback was sought were those who had been involved in the project over an extended period of time and who had developed a shared understanding of the types of tasks. The content focus was the relationship between area and perimeter.

P. Sullivan et al., *Teaching with Tasks for Effective Mathematics Learning*,
Mathematics Teacher Education 9, DOI 10.1007/978-1-4614-4681-1_11,

The Tasks and the Lessons

The following are the three tasks that were provided to the teachers:

Tasks title	Task type	Source
Inside and Outside the Square	Purposeful representational task	Source unknown
Dido's Problem	Contextual task	Adapted from the notes in Curriculum Support for Teaching Mathematics 7–12 [NSW Department of Education, Vol. 13, No. 4 (2008)]
What the L?	Open-ended task	Sullivan and Lilburn (1997)

The tasks were presented to teachers at project professional development sessions for each of three clusters of schools. The teachers worked through each task individually, shared their work with others, and then discussed how their students might respond to each task. The following is a brief discussion of the tasks. The lesson outlines that were provided to the teachers are presented in the following figures.

Inside and Outside the Square

This task provides a set of simple square grid-based shapes that have the same area but different perimeters and another set that have the same perimeter but different areas. The intention is that students work on the workcard with minimal initial input from the teacher (see Fig. 11.1 below). A written student response to the questions "What do you notice?" and "Why might that be?" was expected and intended as the basis of teacher–led discussion.

This has some elements of an open-ended question, but the grid or dot paper is intended to be a representation linking back to the initial purposeful representational task.

Dido's Problem

The intention of this task is for students to observe the way in which a given perimeter can lead to a variety of different areas for a closed shape. As with the other lessons, students experienced area-as-covering and counting units to provide the basis for calculating areas of shapes, and that as a closed shape with fixed perimeter gets "rounder", it increases in area. Figure 11.2 is the handout provided to the teachers.

Inside and Outside the Square

For each of the following shapes find the area and perimeter.

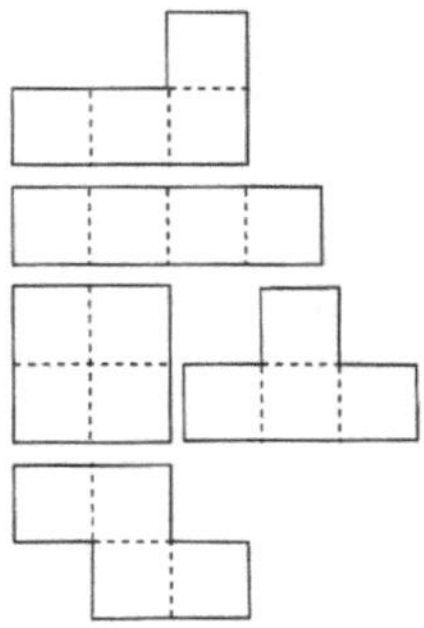

What do you notice? Why might that be?

For each of these shapes find the perimeter and area. What do you notice?

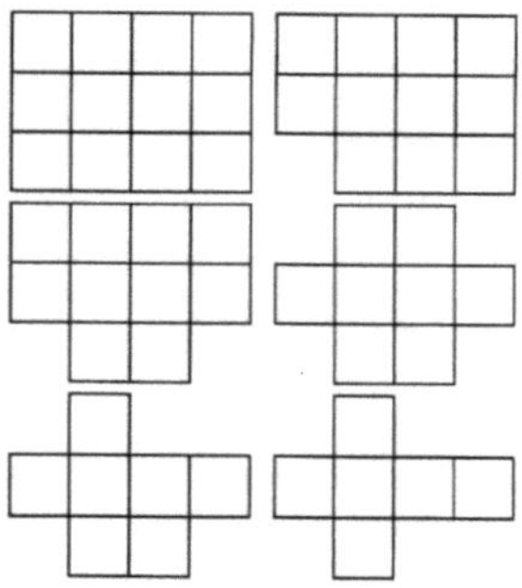

What is the relationship between perimeter and area of these types of shapes? Using grid or dot paper, ask the students to:
(1) create four shapes that have the same area but different perimeters; and,
(2) create four shapes that have the same perimeter but different areas.

Fig. 11.1 Inside and outside the square workcard

What the L?

The goal of the lesson is for students to realise that different shapes can have the same area but different perimeters, and to develop skills in calculating area and perimeter. It begins with a relatively simple open-ended task focusing on area within a square grid. Figure 11.3 shows the handout provided to the teachers.

At the meetings at each cluster, the teachers tried these tasks and discussed the lesson structures suggested. The details of the data collection were presented and the teachers were asked to teach the three lessons consecutively, although in the order of their choosing.

Dido's problem

You need dot paper, minties (enough for 2 per student), a ball of string and rulers

Introduction

Tell students about Dido, founder and first Queen of Carthage. History tells that she arrived on the coast of North Africa, having fled from her brother King Pygmalion. She asked the local inhabitants for a small amount of land for a temporary refuge. They agreed when she asked for only as much land as could be surrounded by an ox hide. She cut the ox hide into fine strips which enabled her to encircle a very large area, which later became the city of Carthage. Today, we are going to investigate the largest area which can be enclosed by a fixed perimeter.

The task of interest today: [We don't have enough ox hides for one between two (!) so …, instead we will be using Mintie wrappers.] What is the longest continuous strip you can make from a Mintie wrapper without tearing it in to more than one piece, and what is the greatest area which can be enclosed by a length of string that long?

Invite students to tear the Mintie wrapper carefully, trying to make it one continuous long strip. Now invite them to measure the length of their streamer (defined as the distance between the start and finish "as the crow flies", so we are not measuring "around the corner."). The students then cut off a piece of string that long.

Now they should take that length of string and form closed shapes with that perimeter on dot paper. For each shape, they should draw it on dot paper, recording the perimeter (the same each time hopefully) and the area, if they can determine it, either by counting squares or some other means.

One way of emphasising how the area can quickly change while maintaining the perimeter is to form a heart shape (noting the area visually) and then flip the heart "open", thus forming a shape with obviously much larger area.

Encourage them to share their findings. In pulling the lesson together, return to the original Dido context, emphasising how large an area could be enclosed by something as small as a torn Mintie wrapper. Mention that it is now not surprising that an ox hide torn into strips could enclose quite a large area (the city of Carthage).

Fig. 11.2 Dido's problem handout

What the L?

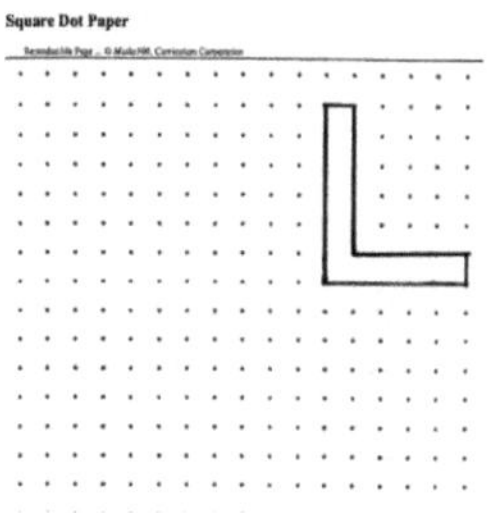

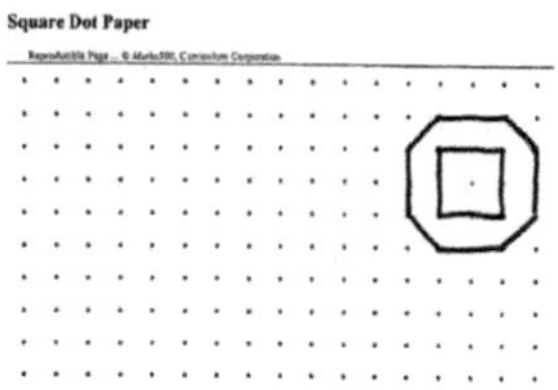

The intention is to allow students to make choices on aspects of perimeter and area. Hopefully the students will experience area-as-covering and counting units to provide the basis for calculating areas of shapes, and they will see in the patterns of responses some insights into the fundamentals of calculating area and perimeter.

There are a number of suggestions in the following, and you can feel free to choose from among them or to do them all.

Part 1: Drawing shapes of a given area

The students are asked to draw, on squared paper, letters of the alphabet using exactly 10 whole squares.

The students can also be asked to draw, on square paper, other letters of the alphabet using a total of 10 squares, using some half squares to make the letters easier to read.

Part 2: Working out dimensions

The goal is for students to realize that different shapes can have the same area but different perimeters, and to develop skills in calculating area and perimeter. Pose a problem like the following:

Imagine that the letter L is drawn on a large piece of squared paper. Its area is 100 cm2 What might be its dimensions?

Teaching approach

Each of the questions can follow the same teaching format.

-First pose and clarify the purpose and goals. The possibility of multiple responses can be discussed.

-Next, students work individually, initially, with the possibility of some group work.

-Then lead a discussion of the responses to the initial task. Students, chosen because of their potential to elaborate key mathematical issues, can be invited to report the outcomes of their own additional explorations.

-Finally summarise the main mathematical ideas.

Fig. 11.3 What the L? handout

Results: Teacher Data

A range of data was collected during the trialling of the three tasks, including

- Teacher questionnaires prior to implementation, including reflections on their planning
- Teacher questionnaires after implementation, including preferences and
- Student questionnaires, including preferences in relation to the different tasks
- The results from the teacher surveys are presented first, followed by the student preferences and then some comparisons. All classes were grade 5, grade 6, or combined grade 5/6

Of the 13 teachers who submitted data related to the teaching of the three lessons, only eight completed both the before and after questionnaire. Their responses will now be discussed.

Order of Teaching

The teachers were asked to indicate what order they intended to teach the lessons and why they chose that particular order. This also corresponded to the actual order taught. Table 11.1 shows the responses.

The choice of the initial task was equally spread between Dido's Problem and Inside and Outside the Square. Teachers who chose to begin with the Dido task either commented that it seemed to "set the scene" or was potentially engaging or interesting for the students. None of the teachers suggested starting with *What the L?* One teacher's justification was that *they appeared to go from "closed" to "open"*. The assumption here seems to be that open-ended questions are harder for the children. While there are some patterns, such as the tendency to teach *What the L?* last, the diversity of responses suggests differences in preferences on the optimal order for teaching.

Previous experience and evidence suggest that many teachers find open-ended tasks difficult to teach, at least initially (see e.g. Anderson, 2003), and assume that they are harder for the children.

One of the teachers who proposed to begin with *Inside and Outside the Square* argued that "I think the understandings and skills required are sequential and this order is best to scaffold the student learning". As a research team, we believe that the content of the three lessons is roughly equivalent, and if anything, the requirement to explain their findings in *Inside and Outside the Square* is quite challenging.

Table 11.1 Teachers' stated preferences for lesson order prior to teaching

Teacher		A	B	C	D	E	F	G	H
Order	Inside	1	1	2	2	2	3	1	1
	Dido	2	2	1	1	1	1	3	3
	What	3	3	3	3	3	2	2	2

	Inside and Outside the Square	Dido's Problem	What the L?
H	With the students' prior experience think this will be easy to teach.	I Guiding the students to make connections or generalising a statement without telling them!	Providing sufficient modelling without "giving it away"
E	Guiding the kids on their discoveries. Giving clear explanations. Asking the best questions. Supporting the kids when they are frustrated. Giving the kids wait time	Bringing ideas together in a summary. Bringing students back to the mathematical focus	Helping the children devise ways of counting squares to find area [Probably relating to the fractions of squares]
A	Helping them make this connection (between area and perimeter)	Getting across conclusively that the perimeter does not need to change for the area to change	Supporting students to use their own strategies for drawing the L without imposing my own way

Fig. 11.4 Sample teacher responses to teaching challenges

We expected that *What the L?* might be an introductory task as it starts with relatively simple square counting. Clearly different people see different things in tasks, and discussion on task potential at teacher learning sessions is valuable.

Teaching and Learning Challenges

Prior to the teaching of the three lessons, the teachers were asked to respond to the questions "What do you expect will be the most challenging aspect of teaching this task?" The responses varied with reference to mathematical as well as pedagogical challenges. They tend to reflect the differences in the type of task. The examples in Fig. 11.4 where each row represents the response of a particular teacher illustrates this (Fig. 11.4).

These responses are thoughtful and insightful and indicate that the teachers are aware of both the potential and the challenges in using such tasks. The language used indicates a focus on guiding and supporting across each of the task types. At the very least, the teachers have learnt the language necessary to communicate about the types of tasks. A common theme in relation to open-ended questions was the acknowledgement of the need to be alert to student strategies and not pre-empt their thinking.

The teachers were also asked "What aspect do you expect the children will find the most challenging aspect of this task?" The responses in Fig. 11.5 are illustrative of the answers provided.

In some cases, the teachers seemed to respond independently of the order they were intending to teach, considering these as individual lessons rather than a sequence. For example, one teacher who was going to do *Dido's Problem* first referred to the challenge for the students undertaking the *Inside and Outside the*

	Inside and Outside	Dido's Problem	What the L?
A	Defining the link between area and perimeter	Tearing the mintie wrappers! Calculating the area of irregular shapes, esp with curved lines – accuracy	I suspect some students will have some difficulty with the concept of 'dimensions'
B	Being able to discuss/articulate what they have found	Understanding that shapes change the area –heart to circle	Halving the squares to make an 'O'

Fig. 11.5 Sample teacher responses in relation to students' challenges

Teacher	A	B	C	D	E	F	G	H
Prefer teaching	What	Dido	Dido	What	Dido /Inside	Dido	Dido	Dido
Student learning	What	Dido	Dido	What	Inside	Dido/What the L		Dido

Fig. 11.6 Teacher preferences following teaching and perceptions of contributions of each task to student learning

Square to be "knowing the difference between area and perimeter and calculating area and perimeter". This would clearly be needed for successful completion of the previous task if the responses were related to their teaching intentions.

While there was some reference to the difficulties of tearing the *Mintie* wrapper, the major challenge identified for *Dido's Problem* was finding the area of irregular shapes. For *What the L?* (abbreviated to "What"), the concept of dimensions, the size of the L, and the counting of squares were identified. The major challenge anticipated for *Inside and Outside the Square* (abbreviated to Inside) was the explanation of the reasoning. These challenges tend to focus on the specific tasks requirements rather than the type of task. While only brief responses were provided, these identify the key aspects of the tasks including the mathematical challenges.

Teacher Preferences

After they had taught each of the lessons, the teachers were asked which lesson did they *prefer teaching* and which did they think contributed most to *student learning.* The teacher responses are presented in Fig. 11.6 in the same order as in Table 11.1.

Even though there was a preference on the part of the teachers for *Dido's Problem* both in terms of their personal teaching preferences and the potential for student learning, the main inference from these responses though is the diversity of their choices.

Prefer teaching	Student learning
Children were keenly involved in the activity. I had to be very specific with my instruction so it was good for me too	For my level 4's the "inside and outside .." was not challenging enough
Loved the Dido problem. Students really got into it. Discussion/ support for each other was great	More linked to real life. Research the story on the internet. Able to visualise area/ perimeter outside
Probably Dido's problem as it was the most open-middle and open-ended. The students responded to it well and experimented with shapes more	It was hands-on with materials and therefore the students could manipulate, test, experiment more. The accompanying story really fired their imagination & motivation as well
I enjoyed Dido's problem. More of an investigation that children could explore	
The story seemed to capture their interest more than others	Dido's problem – because they all started to do a square/ rectangle shape but when encourage to try other shapes, they could see some other shapes had same perimeter as their first– they seemed interested that the love heart had same perimeter but different area

Fig. 11.7 Teacher preferences following teaching and perceptions of contributions of each task to student learning

Dido – I liked the story aspect– kids kept coming back to it as a reference point. Kids could visualise Inside and outside the square – gave the kids lots to talk about. Questions to ask. Most challenging	Most successful way of helping kids learn about area and perimeter. It was a great way to see where their misconceptions were. Visual, labelling drawings good strategy. Lots of discussion between pairs and whole group at different stages, i.e.,during and after activity

Fig. 11.8 Preferences of a teacher who preferred both Dido's problems and Inside the square

The written justifications for those who preferred teaching *Dido's Problem* are presented in Fig. 11.7 with the corresponding response relating to potential for student learning:

Again these are clear, articulate, and thoughtful responses.

One of the teachers identified both *Dido's Problem* and *Inside the Square* as jointly preferred, and the justification is presented in Fig. 11.8:

The justifications of the two teachers who preferred *What the L?* are presented in Fig. 11.9.

Overall, these responses are insightful and reflective, and their preferences are connected to the students' responses. The teachers are clearly interested in students' engagement and have a commitment to student learning.

These responses indicate that teachers recalled aspects of tasks that are engaging and enjoyable for the students. Each teacher's reflection on which task they prefer to teach included reference to the students, indicating the importance of student learning.

The students seemed more engaged with the mathematics in this lesson than the others. This made it easier to get the concepts across	Students all remained on task and engaged, they were discussing the maths throughout. Concepts of area and perimeter consolidated and 'dimensions" introduced
I enjoyed the children challenging themselves with a variety of letters of the alphabet	It's problem solving, trial and error led to a variety of discoveries for themselves

Fig. 11.9 Preferences of teachers who preferred What the L?

There was reference to specific mathematics in the responses justifying the contribution of the tasks to student learning. These responses also indicate the value placed on student participation through reference to materials/hands on/visual, etc. The nature of the content requires a visual focus and this is evident in each of the tasks, so such responses are not surprising. However they do suggest a pedagogical approach where students are physically and cognitively active both in terms of teachers' preferences and their perceived contributions to student learning.

One Teacher's Contrast Between Types of Tasks

The final question on the questionnaire: "What are the key differences between the three task types ?" was intended as a general question focusing on the differences between task types. The teachers' responses tended to focus on the particular tasks separately and provided limited data on the differences. Only one of the teachers provided a response that focused on the different types of tasks. Her responses articulate what she saw as differences between the teaching of the different types of tasks.

With respect to the difficulties of teaching challenges experienced, she responded:

Contextual tasks–Ensuring the focus stays on the mathematics
Open-ended tasks–Making sure students are engaged enough to extend themselves and not taking the easy option
Purposeful tasks–Drawing out the mathematics

With respect to the difficulties of learning challenges students experienced, she responded:

Contextual and Open-ended–Finding strategies for solving the problems presented

With respect to the teaching strategies and pedagogies required, she responded:

Contextual tasks–students' own strategies and processes–they generate the ideas
Open-ended tasks–often more independent, lots of questioning
Purposeful tasks–teacher-centred during discussion–drawing out maths

There is a focus on the student-initiated strategies when discussing both open-ended and contextual tasks. This is consistent with the difficulties identified and contrasts the more explicit focus of purposeful representational tasks. There is recognition of the need to ensure that students engage and extend themselves when using open-ended tasks.

Student Responses

Student survey responses were provided for 227 students across 12 grades. The students were asked to rank each of the lessons in terms of liking, learning, and ease, and explain their reasons. Finally, they were asked what advice they would give to improve these lessons.

Student Preferences

Figure 11.10 presents a graph of students' responses to prompts that invited them to rank the order in which they "liked to do", and to write why they chose that task. The students were also asked to nominate the two tasks from which they "learned the most" and which was "the easiest", using the same format.

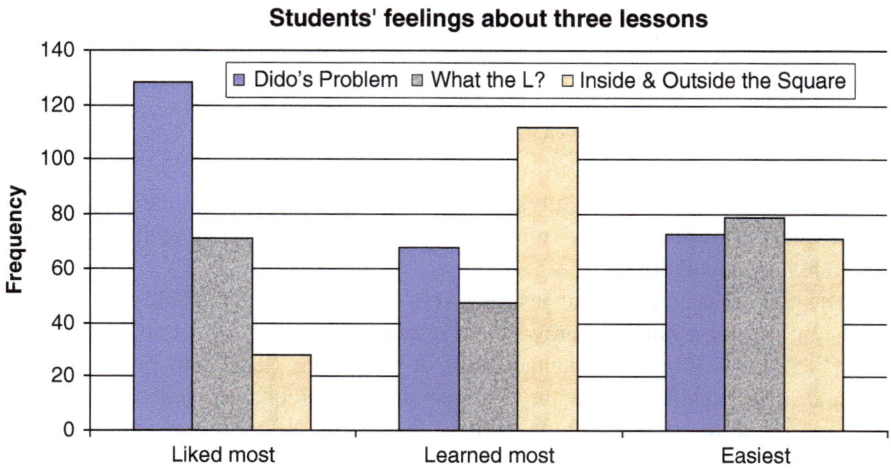

Fig. 11.10 Students' feelings about the three lessons

This indicates that there was a diversity of student preferences, and that they see liking, learning, and easy as different. About half of the students (120) liked *Dido's Problem* the most, followed by *What the L?* (71), and then *Inside and Outside the Square* (28). Even though mentioned by some, we do not think that the sweet (candy) was the reason for the preferences. The students gave meaningful reasons for their preferences.

The main reason students liked the lesson they chose was because it was *fun/interesting/constructive/creative* (112). The next most common reasons were eating the *Mintie* (31), followed by the fact that the lesson was practical (24) and challenged their thinking (22).

The students were also asked to explain their choice of the lesson they learned the most from. The most common response was that they learned most from the lesson *Inside and Outside the Square* (112), followed by *Dido's Problem* (65), and then *What the L?* (47). The main reason for their choice was that it was challenging, including reasons such as *more complex, lots of steps, have to think*, and *learn something new* (155). The next most common reason was that the task was easy to do (16).

Similarly, the students' responses to the prompt about which lesson they found easiest were examined. There was a diversity of preferences in terms of ease of tasks. There was almost equal numbers of students who considered each of the tasks easiest. *What the L?* (79) was chosen as the easiest, followed by *Dido's Problem* (73), and then *Inside and Outside the Square* (71). The main reason given for their choice was that it was easy to understand and not much to do (114). The next most common reason was that the lesson was fun (36).

In all, these responses indicate that students have particular preferences and are prepared to articulate them. It also suggests that students may see liking, learning, and ease as different, suggesting that this could be a productive area of further investigation.

Advice on Improving Lessons

Students were asked for the advice they would give to improve the lessons. The most frequent response was to make the tasks more challenging, including the use of mental arithmetic (52).

The next most common suggestions were to use different strategies (41) such as revision; mental arithmetic; draw or cut instead of tearing; tell the story that came with it (*Dido's Problem*); change order of procedure; start with inside; explain "inside" and "outside" the square; change number of blocks for *What the L?*; use 15 blocks instead of 10; use 20 blocks; make the shapes different sizes; more interactive; more visual; different order of activities; use animals or cars; and more open-ended more hands-on for *Dido's Problem*.

Students wanted the lessons to be more fun with more snacks (30). However, 35 students indicated that no improvement was required.

Summary Comments from the Student Data

The following are some inferences overall from the student responses:

- Most students liked *Dido's Problem* and identified "fun" most commonly as the reason. The "ease of the lesson" seemed not to be an important factor influencing how much they liked the lesson.
- The students claimed to learn most from *Inside and Outside the Square*. They favoured lessons they found challenging (including more complex, lots of steps, have to think, and learn something new), with 155/227 responses being categorised in this way. In contrast to their reasons for liking a lesson, "fun" was not an important factor.
- The students were equally divided as to which of the three lessons was the easiest. For many students, *understanding* and *not much to do* contributed to making a lesson easy.
- The story behind *Dido's Problem*, mentioned in various parts of the survey, helped motivate some students, but this was not the case for large numbers. Students advised teachers to make the lessons more challenging, suggested different strategies and to make them more fun with more snacks. A significant number felt that no improvement was necessary. Making the lessons easier to understand, more interesting, use computers, work in groups, or work outdoors were not considered by many students as necessary to improve the lessons.

Overall the student responses were perceptive and many responses indicated a desire to learn and be challenged. They included an ability to articulate important mathematical concepts and understanding of the purpose behind many of the teachers' actions. They distinguished between learning and liking, which further supports findings from Chap. 10 that these differences are subtle.

Relationship Between the Student Preferences and the Teacher Preferences

The data from students were collected, collated, and analysed at the class level and so the preferences of teachers and students were able to be linked. Table 11.2 presents the teachers' preferences and those of the surveys provided from their own class.

While both teachers overall indicated a preference for *Dido's Problem* (6/8) and the students overall indicated they liked that task (120/227), this does not appear to be a strong link between the preferences of individual teachers and their class. For only four of the eight teachers was their preferred lesson chosen by more than half of their children. Interestingly, teacher E chose both *Dido's Problem* and *Inside and Outside the Square* as her preferred lessons, but 15/23 children liked *What the L?* best.

The link between the lessons from which the teacher thought the students learned the most and the actual student responses was variable. There was no class where

Table 11.2 Relationship between the student and teacher preferences

	Teacher data		Student data					
			Liked the most			Learned the most from		
Teacher	Pref for teaching	Student learning	Inside	Dido	What the L	Inside	Dido	What the L
H	Dido	Dido	1	14	5	5	8	5
A	What the L	What the L	0	20	3	10	3	10
F	Dido	Dido/L	3	5	11	9	7	2
D	What the L	What the L	3	11	9	14	2	7
C	Dido	Dido	4	13	5	12	9	1
E	Dido/Inside	Inside	3	5	15	11	12	0
B	Dido	Dido	1	4	1	2	0	4
G	Dido		3	8	3	15	5	2

the task from which the teachers thought the students learned the most was identified by more than half of the students as the one from which they claimed to learn the most. Teacher F felt that the children learned the least from *Inside and Outside the Square*, but 50 % of the student responses favoured this task for learning.

This seems to be another area of productive future investigation.

Summary

The teacher and student data provide useful insights into their responses to these tasks. The three tasks discussed in this chapter all proved to be worthwhile, and provided a chance for us to consider in some detail the opportunities and challenges offered by different task types, all focused on the same broad content area, namely the relationship between area and perimeter.

In considering the order of tasks, many teachers seemed to be preoccupied with the view that you start with the representational tasks and finish with the open-ended. Contextual tasks can be chosen both as summative and as motivational. If they are chosen as motivational, then it is usual for them to be introduced first. A further question emerging from this analysis is whether the teachers actually considered these as a sequence or as isolated experiences. The focus on lessons or tasks can provide for a disjointed approach. This is explored further in the discussion of sequences in Chap. 8. While our focus is on tasks, we acknowledge the need for coherence and connections across experiences.

As with previous student data, we were impressed by their thoughtful responses and students' apparent ability to distinguish between the constructs of learning, liking, and ease.

The data reported in this chapter support the view that there is a diversity of beliefs across students and teachers in terms of preferences for different types of tasks, in terms of their contribution to motivation, enjoyment, and learning. A theme common to much of this book is that the data indicate the importance of providing a wide range of types of tasks in a variety of different sequences, in order to meet the needs of as many students as possible.

Chapter 12
Conclusion

In the *Task Types in Mathematics Learning* Project and in this book, our goals were to describe

- How different kinds of tasks contribute respectively to mathematics learning
- The features of successful exemplars of different types of tasks
- Constraints teachers might experience when using tasks
- Associated teacher actions that can best support student learning

We developed a strong professional relationship with a small number of committed teachers who helped us to explore these important aspects of the teaching and learning of mathematics. In this chapter, we draw together the key findings from our work on mathematics tasks, as presented in this book.

In summary, it seems that the choice and use of tasks are central to effective mathematics teaching, and that the potential of the learning is connected to characteristics of the task. Some of the key characteristics of tasks that we recommend are that they

- Engage students in doing important mathematics, fostering meaning making, understanding, and connections to other aspects of mathematics
- Are challenging for most of the class, with the pathway to the solution not being obvious to the students
- Require students to think, make decisions, and communicate with each other
- Prompt thinking and reflection
- Use contexts or situations with which the students are familiar and which they see as potentially useful for them or connected to their lives

Of course, there are few tasks that can do all of these things, but the challenge for the teacher is to use these ingredients to create a balanced and healthy diet, even though the individual "meals" may use only a few. We also elaborated key background variables including teachers' knowledge, their beliefs, and constraints they experienced when using different types of tasks.

P. Sullivan et al., *Teaching with Tasks for Effective Mathematics Learning*, Mathematics Teacher Education 9, DOI 10.1007/978-1-4614-4681-1_12,

In particular, our research focused on three different types of tasks:

Purposeful Representational tasks involve a model, example, or explanation that elaborates or exemplifies the mathematics.
Contextual tasks situate mathematics within a contextualised practical problem.
Open-ended tasks are framed as an open-ended questions focusing on specific mathematical content.

Each of these task types was trialled with the teachers focusing on one type of task at a time to build confidence in their use, develop a shared understanding of the nature and focus of the task type, and then contrast the pedagogical and mathematical aspects of their implementation. During the latter stages of the project, teachers were engaged in the development of lesson sequences and in exploring student responses to tasks. The perspective of student preferences and related reactions to tasks was a dimension of the data collection that expanded during the project due to the useful insights it provided.

What We Learnt About the Different Types of Tasks

Our task categories were useful for describing characteristics of tasks, and the three types on which the TTML Project focused each describe tasks with useful purposes. We believe that all three types should be part of teacher planning.

We have presented a number of times to groups of researchers and teachers on this work. The typology has been understood well and enabled discussion in relation to the identification of tasks of each kind. It does not provide a clear categorisation that might be used consistently as the types are not mutually exclusive or inclusive, but it is certainly a helpful tool for generating discussion about the mix of types which teachers currently use, and possible changes in that mix or balance.

We found that teachers improved on all aspects of task use over time, especially those who showed strong commitment to the project, with regular attendance at professional learning sessions and taking opportunities to trial a large variety of tasks with their students. We now discuss some of what was learned about each of the three task types, with further discussion in relation to implementation later in the chapter.

Purposeful Representational Tasks

In conceptualising these tasks, it was not our intent to include routine exercises but rather to highlight the richness that is possible in more traditional tasks that make use of representations and other tools that embody specific mathematical concepts. In the development and trialling of this type of task, there were a number of important findings. We noted that while these tasks are not usually contextualised, there is sometimes a “hook” that helps to engage students. While the mathematics may be explicit, extensive exposition by the teacher is not necessary as the provision of the model or representation enables the students to generate the mathematical ideas and justification. The model, representation, or tool is ideally linked closely to the mathematical concept being developed, in order to be effective.

We noted that with these types of tasks, the mathematic focus is pivotal and it seems that teachers might be less willing to deviate from the mathematical intent than with contextualised tasks, for example. It is also important to recognise the role of the tool or model as an ongoing cognitive support for the student, and its introduction should therefore not be rushed. Some of our teachers during the trialling phase were reminded of the importance and contribution of such tasks to a balanced teaching programme.

Contextual Tasks

There is a strong consensus in the research literature that the nature and quality of student learning are determined by the nature of the task and the way it is used. The greatest gains on performance assessments, including questions that require high levels of mathematical thinking and reasoning, are related to the use of instructional tasks that engage students in mathematics with connection to meaning. Contextual tasks have considerable potential in this respect.

Our claim here is that contextualised tasks can assist students to make connections between mathematics and its applications, and to see how mathematics can help students to make sense of the world, in settings that engage most students. Teachers need to take care that the meaning and realism are not too contrived and that allowances are made for students who might not be familiar with the context. Teachers are also advised to take care that they are not using the context simply as a "hook", which is left behind when the mathematics becomes the focus. In pulling the lesson together, it is important that the teacher focuses on both the mathematics and what has been learned about the context.

Open-Ended Tasks

As with the other types of tasks, open-ended tasks have an important place in mathematics classrooms, particularly for their potential to provide challenge for *all* students. We noted the experiences of some teachers that students are sometimes confronted initially by tasks for which there is a range of possible answers, and even very capable students can be threatened if they have limited experience in working on them. Some teachers also commented that such tasks are not necessarily suited to low-attaining students, but our experience is that, provided the teacher prepares both *enabling prompts* (for students who find it difficult to make a start) and *extending prompts* (for those who finish the initial task quickly), all students can benefit from working on such tasks.

A feature of such tasks is that they can be developed by teachers given appropriate models for task construction, such as the ones we have provided in this book. The experience in the project was that teachers appreciated both the opportunities and the challenges presented when using such tasks, and found that the tasks were both manageable and helpful when used in their classrooms. Such tasks clearly create opportunities for learning mathematics.

Other Types of Tasks

We had intended to include extended investigations as a fourth type of task in our project. This proved to be problematic. One of the major factors was the *time* such tasks take, and the demands of what is seen by many teachers as a *crowded curriculum*, and so the use of such extended tasks was not a common practice amongst the teachers with whom we worked. There was a focus on the lesson as the unit of planning and implementation, and rarely was a given task spread over a number of lessons. The commitment of teachers to more extended investigations in an environment of content-focused planning made trialling of this kind of task difficult from our research perspective.

In conversations with teachers, there was a preference for using purposeful representational tasks to support learning in topics in number and algebra, due in part to the focus on models or representations that are evident in these areas. Many of the purposeful representational tasks involved fractions and proportion reasoning as the mathematical focus. The areas of measurement and statistics were more strongly represented in the contextual tasks that were developed and trialled. In the development of teaching sequences, the teachers identified and used a range of different task types across the quite varied content of data representation, ratio, and financial mathematics. Open-ended tasks were fairly easily created for most aspects of mathematics, and teachers indicated no particular preference for a topic in the development or use of them.

Planning for Task Implementation (or Turning the Task into a Lesson)

One of the components of the framework (Stein, Grover, & Henningsen, 1996) presented in Chap. 3 is the setting up of the task in the classroom by the teacher. This phase of the implementation of the task is a bridge between the intent of the task developer and the student experience. In Chap. 7, we used a framework of teacher actions to discuss the process of turning the task into a lesson and presented a vignette to illustrate possible teacher thinking and planning required in this process.

The framework of teacher actions that we argue can maximise the learning opportunities has the following components:

- Being clear on the mathematical focus and the goals of the lesson for students.
- Considering the background knowledge which students are likely to bring to the task, how to establish this, and likely responses students will make to the tasks.
- Considering ways in which students who have difficulty making a start on the task, and students who solve the task quickly might best be supported.
- Monitoring students' responses to tasks as they work individually or in small groups on the tasks.
- Selecting students who will be invited to share during reflection time.
- Focusing on connections, generalisation, and transfer.
- Considering what the next lesson might look like.

Much of these actions are required prior to the teaching and highlights the need for teachers to be confident and comfortable with the task and aware of its mathematical possibilities prior to implementation.

Differences in Task Implementation

We have noted some differences in the use of the different task types in a lesson. With *purposeful representational tasks*, the mathematics and solution paths are likely to be clearer, with the mathematical focus made explicit. If built around a game, the instructions can be given, possibly with a demonstration game for the whole class to observe. There is also less likelihood of the lesson taking a major detour along the way than for other task types. The reflection on the lesson might take the form of "if you played this game or did a task of this kind again, what would you do differently?" or be written to focus on the individuals student's conceptual development.

A lesson using contextualised tasks will usually start with a discussion of the context, and students' experiences with it or awareness of it. Once this discussion has taken place, the mathematics which can shed light on the problem of interest is then introduced, or the students seek to discover it. The lessons will also generally finish with a return to the context, as well as an overview of the mathematical goal for the lesson. In this way, the context can be used for motivation and engagement and as a means of exploring important mathematical content.

Open-ended tasks are generally of a form that most students can make a start, especially if this type of task is not new to them. Teachers will have considered or prepared enabling and extending prompts for those who need them, and to convince students that they should challenge themselves to move towards generalisation as far as they can. It has been claimed that all mathematics lessons should lead towards generalisation, and this is particularly the case for open-ended tasks, in our opinion.

We have seen that whatever the kind of task(s) chosen for a lesson, the teacher needs to probe students' thinking without extensive "telling", and gather a sense of how individuals and groups are responding to the task(s). We reiterate here a key finding that the teacher's role, carried out effectively, involves many dimensions, much preparation, and a lot of thinking on the run.

In considering what we might look for in classroom observations, a rubric was developed for evaluating the extent of presence of different classroom pedagogies. The pedagogical practices that were the focus of the particular scales that were developed to rate teacher pedagogies across the different tasks types were as follows:

Focus on student thinking. The extent to which students manipulate information by combining facts and ideas in order to synthesise, generalise, explain, hypothesise, or arrive at some conclusion or interpretation.

Making connections. The extent to which explanations and tasks are connected to previous knowledge, indicate the purpose of knowledge, and emphasise both how and why.

Challenging tasks. The extent to which students are presented with tasks which they have not been previously shown how to solve.

Openness. The extent to which the tasks have multiple possible pathways to the solutions and/or multiple possible responses.

Explicit expectations. The extent to which the teacher communicates to the student on forms of communication, solution forms, and nature of expected solutions.

Inclusivity. The extent to which the teacher chooses tasks and gives explanations that are accessible to all students, and the extent to which the teacher seeks to include all students in the classroom processes.

Application and transfer. The extent to which connections are made to practical application of knowledge and the potential for transferring learning to other contexts.

Nature of explanation. The extent to which the teacher allows students to make decisions on strategies and solutions.

Questioning. The extent to which the teacher poses relevant and engaging questions and builds on the responses.

Pulling the lesson together. The extent to which the teacher pulls together key ideas out of the students' experiences with the task during whole class, group, or individual discussions.

Taking opportunities. The extent to which the teacher responds appropriately to particular decision points in the lesson.

Explicit goals for student learning. The extent to which the teacher makes the focus of the lesson and the learning goals for the students clear.

Maintaining the challenge. The extent to which the teacher resists attempts by the students to have the degree of risk reduced by explicit directions.

While it is not expected that all of these would be evident in all lessons, there were some differences evident between the types of tasks. The rubric provided limited data within the project but gives a research-based profile of pedagogical practices that we value.

Opportunities and Constraints in Task Use

Teachers were able to articulate opportunities and constraints in their use of particular tasks, although there was not always consensus in relation to constraints. The opportunities for choice that are provided through open-ended questions were commonly articulated as were the motivation and engagement provided by contextual tasks. The opportunity to link to real-world contexts was strongly identified with contextual tasks. As previously mentioned, a number of teachers commented on the value of purposeful representational tasks and the need to ensure they are part of the mix of tasks.

Constraints were often external to the classroom, including the time required for preparation and planning, the demands of other curriculum areas in the case of primary teachers, and equipment or material requirements. These were common messages from busy teachers in schools where resources are limited. Constraints associated with a teacher's own mathematical knowledge did not tend to be articulated in direct survey questions but were evident through discussion, observation, and response to seemingly unrelated questions.

Some of the teachers with whom we worked had difficulty with

- Seeing the mathematics and learning potential in tasks
- Turning tasks into worthwhile lessons
- Maintaining the challenge of tasks, rather than "watering them down" in response to students' reactions
- Anticipating what students might do with particular tasks and strategies they might use
- Understanding novel strategies developed by students as they solved problems
- Managing end of lesson reviews and
- Adapting task types to new content when creating tasks for themselves

Limitations in teachers' knowledge are clearly related to these difficulties and can be seen as constraints to the successful implementation of quality tasks. From our experience, it is important to provide as many opportunities as possible for teachers to work through new tasks in the role of students, allowing relevant content and pedagogical content issues to arise and be addressed.

It has been argued that teacher beliefs also have an impact on their implementation of tasks. If a teacher sees mathematics as a bag of tools, as mechanistic (Ernest, 1988), then they may see limited value in the open-endedness of tasks and limit the challenge and choice. There is evidence that teacher beliefs influence their choice of tasks as well as the order in which they use particular task types.

In evaluating tasks and lessons, teachers often focused initially on student enjoyment and engagement rather than the mathematics being learned. They talked about a particular game being enjoyable or a context providing interest to the children, but they rarely discussed students and their motivation in relation to the mathematics. We seemed to see a shift as the project progressed to a more nuanced response that included both mathematical and pedagogical insights, particularly amongst the more committed and engaged teachers.

The challenges that teachers identified were dependent to some extent on the tasks type. The following quotes from the teachers exemplify the pedagogical challenges of the different types of tasks:

> Purposeful representational—"Guiding the kids on their discoveries". "Giving the kids wait time"
> Contextualised—"Guiding the students to make connections or generalising a statement without telling them..." "Bringing students back to the mathematical focus"
> Open ended—"Providing sufficient modelling without 'giving it away'". "Supporting students to use their own strategies"

Teachers varied in their preference for particular types of tasks, the students with whom they believed they were best used, the perceived difficulty in creating them, and the order in which they might use them in a particular unit of work. Some teachers, for example, indicated that they would usually start with purposeful, representational tasks, before moving on to "more challenging" open-ended tasks, with contextualised tasks seen as applications, *once the mathematics was understood.* Others saw contextualised tasks as the starting point, because they provided students with a rationale for their learning.

We noted that sometimes teachers would pose tasks to students without having worked on the problems themselves. While there are times when the teacher genuinely modelling the processes of problem solving students go through might have advantages, because they have never worked through a task of this kind before we believe that this is rarely desirable. If a teacher has not worked through the mathematics in a given task, and preferably had the opportunity to discuss with colleagues both the mathematics in the task and possibly directions the lesson might go, then we see that teacher as not adequately prepared for the work of teaching. Of course, it is difficult to anticipate exactly how a given lesson might go, but a teacher who has thought through the mathematics of a task and its implementation will be well placed to respond to issues and detours that might arise during a lesson.

The mathematical knowledge of teachers, including specialised content knowledge, horizon knowledge, curricular knowledge, and knowledge of students, clearly influences their choices of tasks and the way they use them. There are two aspects to this. First, mathematics education programmes for both prospective and practising teachers need to identify key aspects of the specialised mathematics that teachers require and provide those teachers with appropriate experiences. For example, teachers should know that many practical problems can be solved using intuitive approaches; different people use different approaches and many of these approaches are valid. Teachers should be exposed to a variety of such problems, should see approaches to managing the diversity of strategies modelled, and have opportunities to discuss the importance of openness to understanding intuitive suggestions from students. Second, it has to be accepted that it cannot be expected that teacher of mathematics know already all of the mathematics they will meet in their professional lives. It is critical, however, that all teachers are aware of any limitations in their mathematical knowledge and know how to find out the mathematics they need when they need it. This can even include saying to the class "That seems interesting but I am not sure I understand. Let us discuss this again tomorrow".

Students and Mathematics Tasks

Possibly the most insightful aspects of our findings from this project related to our findings about *students* and tasks.

Students were able to distinguish between learning and enjoyment. When asked to rate particular tasks within units of work, students could clearly distinguish between tasks in relation to the level of enjoyment involved, and the contribution that a given task made to their learning of the relevant content. We had anticipated that rating tasks in terms of enjoyment would be straightforward for them, but we were delighted that they were able to also judge a task for its learning potential.

Students were able to articulate their preferences for tasks and lessons in sophisticated and subtle ways, although these varied substantially, suggesting that variety in both task type and lesson structure is critical if the learning and enjoyment needs of all students are to be accommodated. Students demonstrated a range of levels of satisfaction and confidence at all grades from 5 to 8, with little difference across the 4 years. We anticipated that students might be less satisfied and confident as the years progressed, but this was not evident in our data. A key recommendation is that teachers should find out about the levels of confidence and satisfaction for all their students.

In describing their ideal lesson, students were articulate, with interesting variations between those who commented on features of content, and those whose interest was in pedagogy. Many students unprompted stated that they like to be challenged. There is a perception that students in the middle years of schooling are looking for "an easy life", and while this might be true for some, the number of students seeking real challenge in mathematics classrooms was encouraging.

The data indicate that in all classes, there is a range of levels of student confidence and satisfaction, and that the key variability is between students rather than between classes. A common finding across the different data that were collected is that there was a diversity of students' responses to the types of tasks that students claim to enjoy and from which they report they can learn.

The responses from students' unstructured reactions to lessons were similar to the above. While some students commented on pedagogical aspects of lessons, others focused on the content. The method of grouping was important for nearly all students, as were the quality and timing of teacher interactions with them. The tension mentioned above between their liking of an experience and the extent to which they feel it helps them learn was also prominent. There is a need for teachers to be explicit about each of these aspects. Again the importance of all teachers adopting a variety of approaches is emphasised.

We set out to use particular types of tasks as a way of exploring mathematical task implementation by teachers and students. The nature and form of tasks are illustrated through a discussion of actual tasks, their mathematical potential, and the experiences of their use by teachers. In Chap. 13, we elaborate five tasks of each task type, discussing the mathematics potential in each, and reporting on the experience of using them within the project.

Chapter 13
A Selection of Mathematical Tasks

In earlier chapters, we reported a range of experiences from the use of tasks in the Task Types and Mathematics Learning Project. In this chapter, we present an outline of 15 tasks. The structure of these tasks is not intended as a recipe for their use by teachers and teacher educators, but is a collection of insights from their use, highlighting some of the aspects which a teacher might consider when using them.

The tasks have been chosen because we believe they share the following features:

- They address important mathematics in the middle years of schooling.
- They could be described as purposeful, representational and/or contextual, and/or open-ended tasks.
- They have received positive feedback from teachers during trialling, indicating that they engage students.
- They provide the opportunity to learn more about what students know and can do.
- They cover a range of mathematical content across Number, Geometry, Measurement, and Probability and Statistics.
- They can successfully be undertaken by students using a range of approaches.
- They require students to think about strategies for solving problems.

In the description of each task, we include the following, as appropriate:

- A statement of the task and the source
- An outline of the main mathematical focus of the task (level, topic, and main intended activity, e.g. building understanding and reasoning)
- Information relevant for teachers such as
  - Insights from teaching
  - Solution methods
  - Potential student difficulties
  - Suggestions of enabling and extending prompts, for students who may have difficulty starting the task and students who find the task straightforward, respectively

P. Sullivan et al., *Teaching with Tasks for Effective Mathematics Learning*, Mathematics Teacher Education 9, DOI 10.1007/978-1-4614-4681-1_13,

- Some student work samples collected during trialling
- A suggestion of what might prove useful in the following lesson

The tasks are chosen because they were selected for use by the project teachers and because they exemplify the characteristics of the respective task types.

Purposeful Representational Tasks

The following five detailed examples of purposeful representational tasks were developed or adapted during the *Task Types in Mathematics Learning* project. They involve a range of content from the middle years and also provide examples from different categories of this broader task type.

The five tasks are now listed, with brief background information:

Chocolate blocks. As mentioned in Chap. 4, this was one of the more popular tasks identified by teacher in the first phase of the project. While it starts with a context of sharing chocolate blocks, the model of the block and the division as sharing provides the explicit mathematical focus. It proved to be a powerful and engaging task.

Clues on cards. This task uses a set of carefully designed cards that a group of students use to answer a question or solve a problem. These cards are modelled on those developed by the Lawrence Hall of Science (e.g. Erickson, 1989). It was one of the tasks that the students in the lesson sequence described in Chap. 8 identified as one they learned the most from. The particular content focus is on measures of central tendency.

Decimal maze. This task takes the form of a game and challenges the common misconception that multiplication makes a number larger. Pairs of students use calculators to keep score as they move around a carefully designed playing board. Extensive mental computation or estimation is required for success. In the initial trialling of tasks, this was one of the most popular.

Smartie prediction. This task engages students in the representation of information in idiosyncratic but purposeful ways. Students are challenged to represent the contents of a packet of smarties in a form that will enable them to keep track of how many of each colour are left as smarties are taken out of the packet one at a time. It uses a game or competition context where students are awarded if they correctly predict the colour of the next smartie to be removed. The mathematics focus is on probability including both language and simple calculations.

Matching graphical representations. The use of tasks that require the matching of different mathematical representations was an important category within this task type. The matching process provides opportunities for comparison to clarify the specific mathematics that is represented as well as opportunities for engaged practice. In this particular task, the same set of data is represented using column graphs, pie graphs, and box and whisker plots. A careful choice of the data enables the specific mathematical features of the representations to be visible and provides potential for rich discussion between students.

Chocolate Blocks

Three chairs are set in the front of the classroom with chocolate blocks placed on each of them. On the first chair is one block, the second has two blocks on it, and the third has three.

Ten students leave the room, and one by one come back in. As they return, they are asked to decide which chair they should stand behind so that they get the most amount of chocolate when it is divided up equally among those standing behind each chair at the end of the activity. Before each of the final two students re-enter and make their choice of where to stand, the class, as a group, discusses where they think they each of the last two should go, and why.

Source

Adapted from Clarke, D. M. (2006). Fractions as division: The forgotten notion? *Australian Primary Mathematics Classroom*, *11*(3), 4–10, who in turn adapted it from the work of Malcolm Swan in National Curriculum Council. (1991). *Mathematics programme of study: Inset for key stages 3 and 4*. York, UK: Author.

Mathematical Focus

This task is designed to target development of students' understanding that fractions are not just "parts of a whole", but can also be considered as division, or as a quotient. The relative size of fractions is also a key focus.

Insights from Teaching

Visual Display

A powerful visual demonstration is obtained if each group lifts up their trays with the chocolate on them. For a group with (say) five students and three blocks of chocolate, the fraction is formed by three blocks on the tray top, the tray as the vinculum of the fraction (the dividing line) and five people underneath—3/5, 3 shared between 5!

Debrief

Some questions that can be discussed as a whole group:

If, at the end, you had the choice to move to a different chair, would you do so?
Where would you choose to stand in the queue? Is it best to go first or last?
What strategies would you use if you were in the line?

Look at the following examples after the 9th student has made a commitment. Decide where you would go and how much chocolate you would get if you were the 10th student, assuming you are trying to maximise your own amount of chocolate.

No. of choc blocks	3	2	1
No. of Students	4	4	1
	4	3	2
	6	2	1
	5	2	2
	5	3	1

How can we arrange it so that the 10th person gets the same amount no matter where they chose to stand?

(N.B. Shaded row in the table above gives the 10th student the same outcome regardless of chair chosen).

Extension Suggestions

More capable students could be challenged to investigate extreme cases. For example, can they create a scenario where the 10th person gets the greatest, or the least, amount of chocolate possible? Students could also be challenged to devise a scenario where everyone at all chairs gets the same amount of chocolate—is this possible?

The Next Lesson

Discuss or act out the following situations:

Where the chocolate is out of the packet and clearly already subdivided. In this discrete case, the students are now using the fraction as an operator notion, as they calculate, say, 1/3 of 24 blocks, or 2/5 of 20.

Where there are more blocks of chocolate at a chair than people. In this case, the context enables a discussion of improper fractions, where the mixed number equivalent is either obvious or can easily be determined.

To increase the level of difficulty, the number of blocks on each chair could be varied, for example 5 on one chair, 4 on the next, and 3 on the last. Other alternatives might be increasing the number of chairs and also the number of blocks, or varying the number of students sharing the blocks.

Clues on Cards

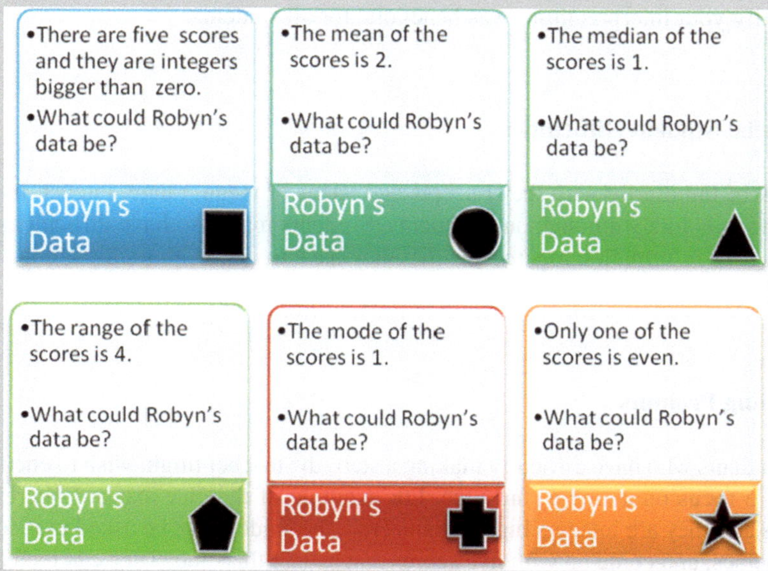

In small groups, students collectively reconstruct data from a set of clue cards based on an understanding of the terms *mean*, *mode*, *median*, *range*, *integer*, and *frequency*.

Source

Cards sourced from Gould, P. (1993). *Cooperative problem solving in mathematics*. Sydney: Mathematical Association of N.S.W.

Mathematical Focus

This task uses the process of collaboration and the sharing of information to answer a mathematical question. The process of interpreting clues and working backwards to generate the data requires skills in interpreting data. Understanding of the concepts of mean, mode, median, range, and frequency is required.

Insights from Teaching

Have a mathematics dictionary on hand for students to look up the meanings of forgotten terms during the reconstruction of data phase.

The teachers who trialled this task found that it was important to ensure that you have students with a range of capabilities in each group. Students who are having difficulty accessing the task will benefit from the input of other students. The need to justify your interpretations was beneficial for all students.

Potential Student Difficulties

A lack of familiarity with the concepts and terms of mean, range, integer, median, and mode is likely among some students. Students may also have difficulties with piecing the information together, in that they have to deal with multiple constraints at once.

Enabling Prompts

For students who have difficulty making a start, the teacher might wish to encourage them to focus on a single card clue. For example, if the clue states: "There are 5 scores and they are integers bigger than zero", the student can be asked to ignore all other clues, and come up with five scores that satisfy this clue. They can then add in another clue and see the way the possibilities narrow.

Another student might be told not worry about the clues initially but simply be asked to "show me some scores which have a range of 6".

Extension Suggestions

Students use previously developed data, such as sporting scores or weather temperatures over time, to develop a set of clue cards. These new clue cards are given to a different group of students to solve.

Students create a "mystery story" to match the reconstructed data. Students aim to incorporate the clues within the text of the story to help others piece the mystery together.

Student Work Sample

The work sample below shows the response of a group of students to a different set of cards that are shown below.

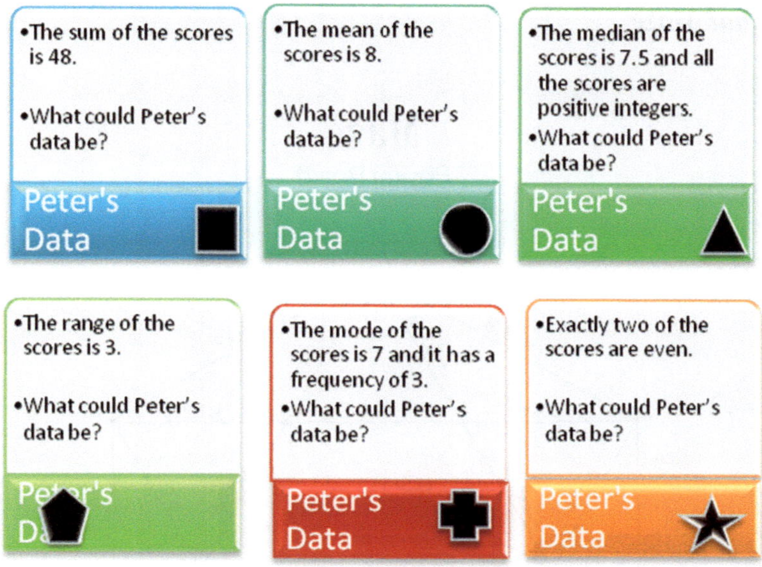

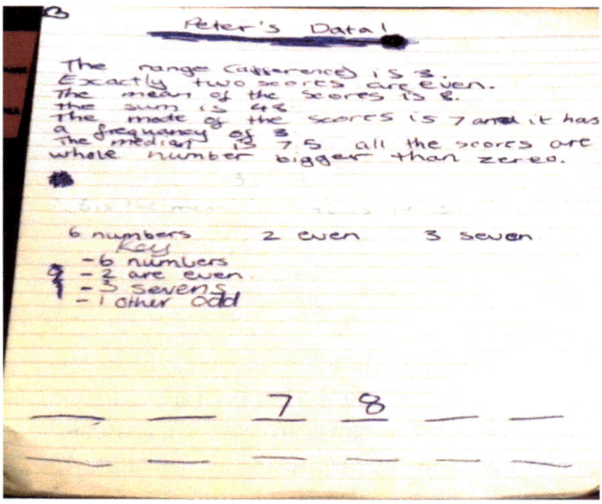

This work sample shows that students are working systematically through the clues to reconstruct Peter's data. The students have demonstrated an understanding of the terminology in this task, e.g. median—the students placed 7 and 8 in the central positions to produce a median of 7.5.

The Next Lesson

Students create their own set of cards, quite similar to the ones used in the lesson, once again involving range, mean, median, and mode. Note that this is closer to what the students have experienced and should be more straightforward than the extensions above, and therefore appropriate for all the class.

Decimal Maze

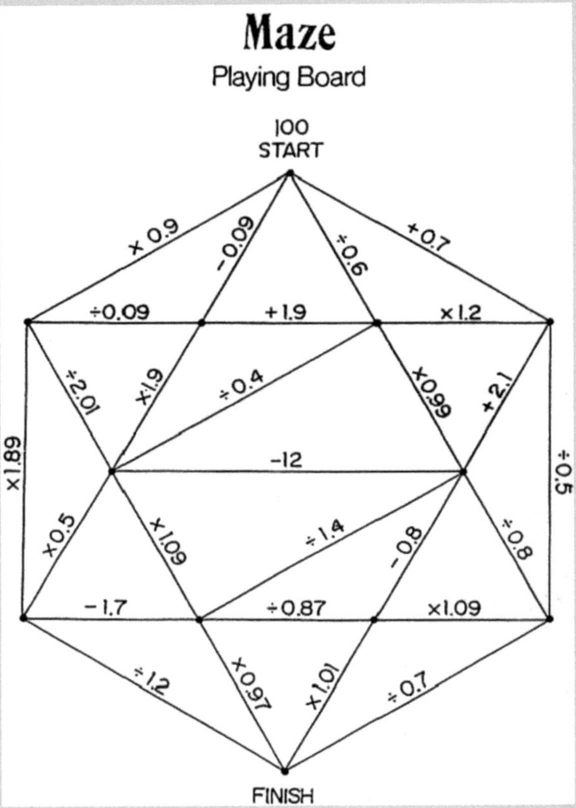

The game is played in pairs. Each student has a calculator, taking turns moving a single counter around a maze board. They each start with the number 100 and are given a choice of alternate moves to make with associated mathematical operations to perform on the number (using a calculator). A separate running total is maintained on each student's calculator throughout the game. The player with the lowest number when the counter reaches the finish is the winner. In playing this game, the common misconception that multiplying always makes a number bigger and dividing always makes it smaller is challenged.

This activity helps students to develop number sense of the four key operations (+ − × ÷) when applied to decimals, in an engaging and thought-provoking board game.

Source

This is a well-known task available through NCTM.

Materials

A game board for each pair
A calculator for each player
A counter for each pair

Mathematical Focus

The key mathematical focus is that multiplying by a number less than 1 gives a smaller answer and dividing by a number less than 1 gives a larger answer.

Computational estimation of division, multiplication, addition, and subtraction of decimal numbers is required. It is likely that some students may have helpful ways of thinking about multiplication and division conceptually (e.g. "100, how many 0.6s?" or 9/10 of 100").

Students can also develop strategic skills by looking ahead to future moves that might maximise their chance of winning.

Insights from Teaching

The game takes approximately 15 min to play, but there is considerable value in playing it a number of times. It consolidates the ideas learned and can be done during the same session or returned to later.

Potential Student Difficulties

Multiplication and division of decimals are challenging for most students in the middle years. Some teachers found that drawing diagrams of division by decimals (e.g. $6 \div 0.5$) and using fraction language like "how many halves are there in 6?" were needed to help students understand the concepts and make reasonable estimates.

Students are able to engage in the game but not always the mathematics. It is less a case of enabling prompts in this particular case than strategic questioning and intervention. The process of estimating and then using the calculator provides immediate feedback to the student and is often a teachable moment.

Extension Suggestions

Students can be encouraged to change the rules to the larger number wins or to consider their partner's next move and be strategic. Students can also be encouraged to reflect on what they have learned. Some examples of student reflection are included below.

Student Work Samples

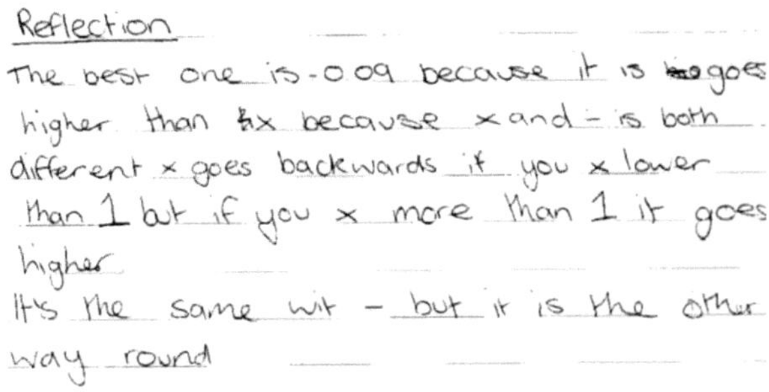
Reflection
The best one is ÷0.09 because it is goes higher than x because x and ÷ is both different x goes backwards if you x lower than 1 but if you x more than 1 it goes higher
It's the same wit ÷ but it is the other way round

This student has identified that dividing by 0.09 results in the largest increase in a number on the game board and that both multiplying and dividing can enlarge or reduce a number. He/she has recognised that it is important whether the number is greater or less than one when determining if the result of multiplying or dividing is an increase or a decrease.

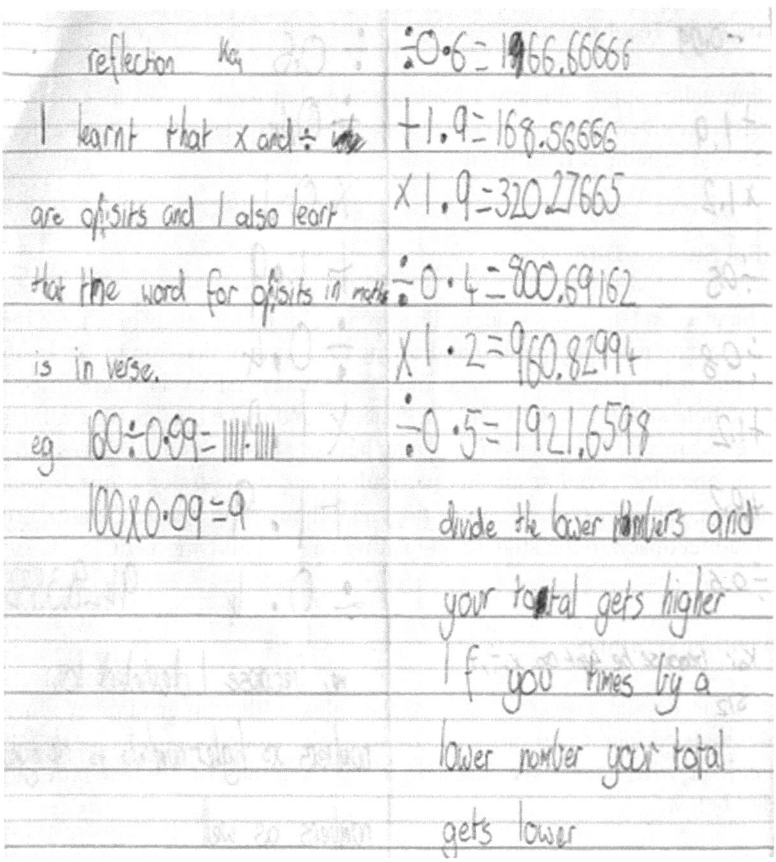
reflection
I learnt that x and ÷ are opisits and I also learnt that the word for opisits in maths is in verse.
eg 100÷0.09=1111.1111
100x0.09=9

÷0.6=1966.66666
÷1.9=168.56666
x1.9=320.27665
÷0.4=800.69162
x1.2=960.82994
÷0.5=1921.6598
divide the lower numbers and your total gets higher
If you times by a lower number your total gets lower

This student has clearly identified that multiplication and division are inverse functions. He/she understands that multiplying does not always increase a number, and dividing does not always decrease it.

The Next Lesson

This task has a very specific focus and its sequencing in a teaching programme can vary. It could be included not only in a unit of work focusing on multiplication of decimals or fractions but also as a standalone task.

Smartie Predictions

Children are shown the contents of a small packet of *Smarties* (or M&M's) and are asked to represent them on paper. The Smarties are put into a bag and taken out one at a time. The children make a written prediction of what colour they think will come out next. They begin with a score equal to the number of Smarties originally in the packet and get a point if their prediction is correct and lose a point if it is incorrect. The winner is the child with the highest score when no smarties are left in the bag.

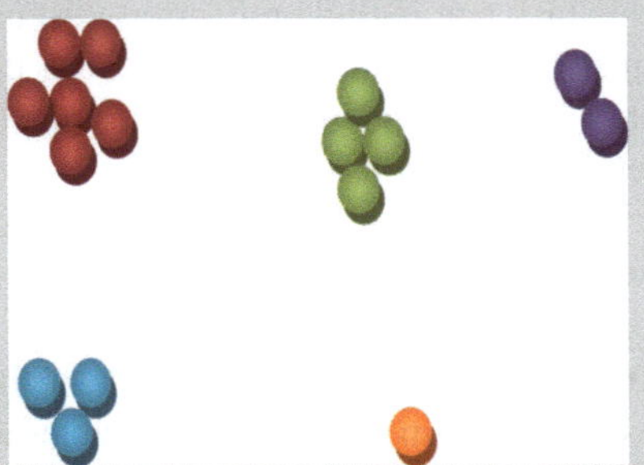

Source

We first became aware of this task through the *PriME* project in the UK. This task has been adapted from the way it was used in that project.

Mathematical Focus

A basic understanding of the outcome of chance events (likely, unlikely, and impossible) is necessary, although this notion is further developed throughout this task. The task provides opportunities for development of chance language such as prediction, possible, certain, impossible, and likely. It builds on a basic understanding of the outcome of chance events and provides an opportunity for estimation of probability based on proportions of Smarties remaining in the bag.

In addition, the systematic recording of information both before and during the game is an important focus. It is likely that there will be many different ways of successfully recording the outcomes and these may lead to valuable observations, insights, and discussions.

Insights from Teaching

As an introduction, the teacher might pose some initial questions using two different coloured counters and a bag. They might place ten counters of the same colour in the bag and ask the children to predict what colour they think the counter would be if we took one out. Then try it with 9 and 1 and then 5 and 5. Depending on the students' answers, this can be explored a little further. Some grade 6 students in the trial were using percentage language and were able to allocate an accurate probability to this simple situation.

To assist the children in making their prediction, they are asked to represent the colours on a piece of paper. This is to enable them to keep a record of which Smarties are left. Some group discussion of possible strategies can be included, but the children's varied strategies are to be encouraged.

The debriefing of this task can include reflection, discussion, or sharing on the success of the representations used and the strategies for deciding which colour is most likely to be selected next. Playing the game again encourages students to look for a more efficient method of recording.

There are a couple of decision points or variations possible. If the children are reasonably confident with their explanation of probability in the introductory session, then on the second playing of the game you might ask them to record the chance of their prediction as a percentage or a fraction. It is not the intention to make this a major focus, but to begin to assign some values at selected intervals; e.g. if there are 6 left and 3 are pink and you choose pink, what is the chance of drawing a pink? For an older class (grade 7 or 8), this might be an important feature and expectation.

In the trial, it was also found to be useful for students to share their varied recording styles, for other students to evaluate and reflect upon. It is unlikely that the higher scores of students means a better recording system, as there is a large chance component with this activity and so identifying successful strategies can be a challenge. The teachers used observation to identify interesting and potentially useful ways of representing the information and determining who should share. The student work below identifies some of the different methods that students used.

There are some advantages in pre-arranging the contents of the packet. When there is a large proportion of one or two colours, there is more chance of the students being successful in their predictions.

Potential Student Difficulties

In trying this task with younger children, a number had difficulties keeping track of the number of Smarties and colours which had already been drawn. This is not a major problem, as the intention is for students to develop skills to manage this for themselves, and the teacher should resist intervening in the first game. The opportunity for reflection and sharing enables these children to be more successful in the second game.

The small probability of the events involved means that even when children make appropriate probabilistic decisions, they are not always successful. This is an important lesson but a frustrating one. In the trial, one child predicted brown (the most probable colour) for two successive turns and then changed due to "frustration" only to have a brown selected on the next turn.

Student Work Samples

Second game of "Smartie Predictions": Example One

This student has devised a logical recording system for the number of smarties in the bag after each draw, and recorded the cumulative points scored as the game progressed. However, number of Smarties drawn is not the same as the scores shown. He/she has not demonstrated an understanding of how to determine the most likely smartie colour to be drawn next, as can be seen from the above sample of work. The student also comments that he/she will do better at the game in future, but does not know why—a further indication that the concepts of probability are not fully understood.

Second game of "Smartie Predictions": Example Two

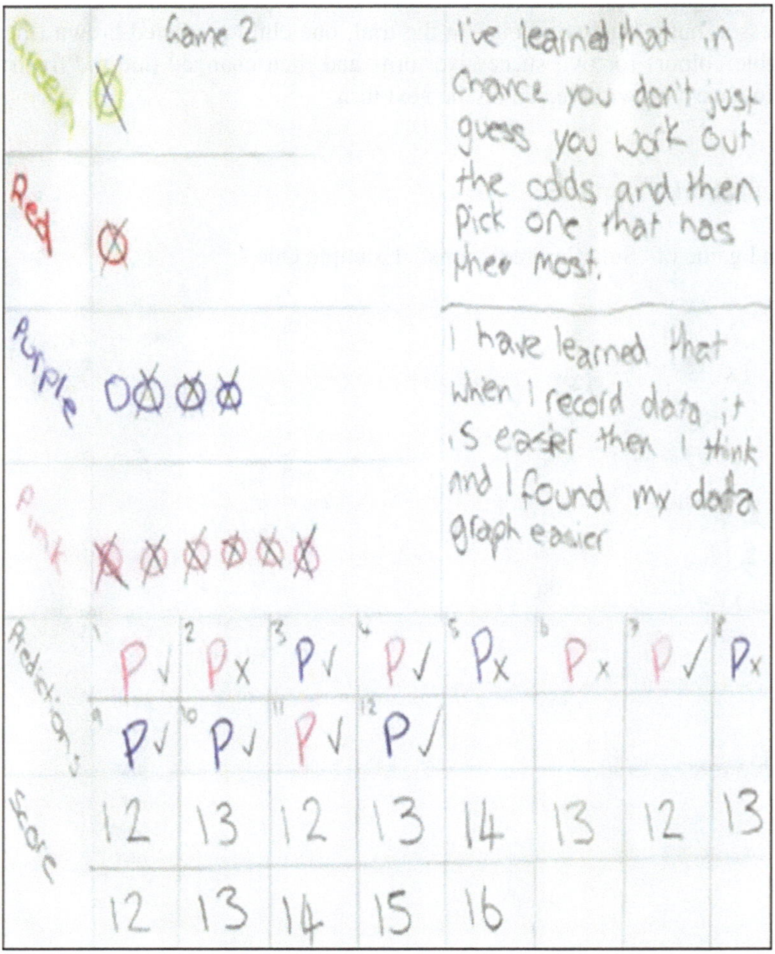

This student has devised a logical recording system for the number of Smarties in the bag after each draw, and recorded the cumulative points scored as the game progressed. He/she has also reflected upon the usefulness of systematic record keeping. From the selections made by the student and the post-game reflections, it can be seen that this student has a good understanding of which colour Smartie is most likely to be selected on each draw and therefore, early principles of chance and probability.

Matching Graphical Representations

Students are provided with cards of box and whisker plots, pie charts, and bar graphs, and are asked to match equivalent cards created from the same sets of data. Working in pairs, students are engaged in a qualitative analysis, unpacking the information contained within each of the graph types and then considering the benefits and uses of that representation.

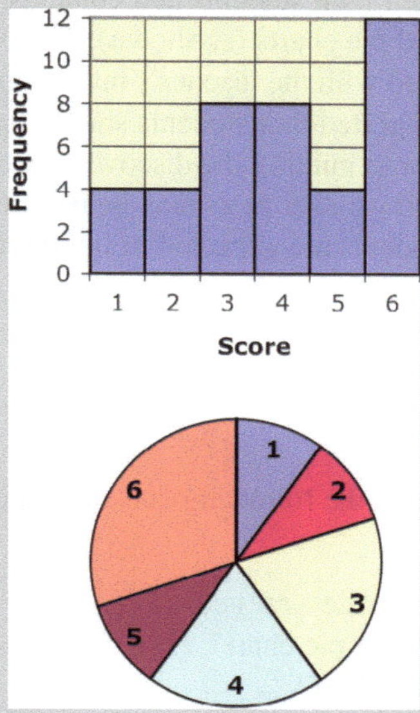

Source

Activity and cards © National Centre for Excellence in the Teaching of Mathematics (UK).

Materials

For each pair of students, you will need a set of each type of graph.

Mathematical Focus

In analysing multiple representations of the same data, students develop a deeper understanding of statistical measures. Students use qualitative interpretation of pie charts, bar graphs, and box and whisker plots. They also gain an understanding of the specific data contained within the different chart types, and the usefulness of these representations.

Insights from Teaching

Most teachers provided some introduction to the different representations. An overhead or data projector was found to be useful during the introduction for display of graphs.

A detailed description of how to use this activity was provided by NCETM, but teachers adapted this to their own students and contexts. For younger students, the focus was on bar charts and pie charts (as shown). The box and whisker plots were introduced later, if this fitted with the teacher's intentions.

Teachers in the trial suggested that students should have some understanding of box and whisker plots prior to guiding the discovery of links with other chart types in this activity. The box and whisker plots may be better suited to a subsequent session in grade 7 or 8, as students are expected to interpret and use this type of chart at that level.

Possible Enabling Prompts

When learners are struggling, the following questions might help them to develop a strategy.

Which bar charts have the smallest range?
How is the range shown on the pie chart?
What is the mode on the bar chart?
Which pie charts have the same mode?

Further questions that will support the statistical concepts for the box and whisker plots include the following:

Can you sort the cards into those that have a large range and those that have a small range?
Can you sort the cards into those that have a large median and a small median?
Does the distribution look spread out (the difference between the smallest value and greatest value is large), or is it concentrated in a few scores (the difference between the smallest value and largest value is small)? Does the distribution look symmetrical, or is it skewed?

Extension Suggestions

The original source material provided a set of "Making your own cards". These can provide extension when either two or three of the different representations are provided.

Alternatively, if the students have access to Excel, they could create their own series of graphs and share these with other groups in the classroom.

The Next Lesson

The students could work on a task such as Seven People went Fishing (Seven people went fishing. The mean number of fish caught was 7, the median 6, and the mode was 5. How many fish might have been caught by each person?).

Alternatively, they could focus on one of the representations and collect and represent data in a context of interest to them, in that particular representation.

Five Contextualised Tasks

In this section, we provide five detailed examples of contextualised tasks which were developed during the Task Types in mathematics learning project, used extensively within the professional learning sessions, used by teachers in classrooms, or a combination of these. All tasks involve starting with a "real" context hopefully of interest to students, and addressing relevant middle years mathematics using the context as something of a "springboard" into mathematics. In every case, our hope is that students and teachers will learn worthwhile and important mathematics, while at the same time learning more about the context involved, and seeing some ways in which mathematics can help us to make sense of the world.

The five tasks are now listed, with brief background information:

Music cards. It is often difficult to find tasks which focus on fractional/proportional reasoning, involving contexts of interest to students. Anne Roche (Australian Catholic University) developed this task around the context of "music cards", which enable individuals to download songs from the Internet for a fee. In this case, students justify a choice between two kinds of cards, one for which they can download 16 songs for $24 or 12 songs for $20, respectively. The focus here is proportional reasoning and comparative value.

Mike and his numbers. Mike was watching the television one night but was bored. He started writing all the numbers from one to one million. After nearly 3 years, he had succeeded. Students are challenged to determine the total number of digits which Mike must have written if he wrote all these numbers. They justify their answers comparing these to the incorrect answer in the newspaper article which reported Mike's achievements. The focus here is number patterns.

Comparing coins from different countries. Students' awareness of different countries, their currencies, and their relative values is enhanced as they work on this task. Each student is given a coin or a number of coins from a particular country, and they are invited to investigate the country of origin of the currency, the coins' value in

Australian dollars, and decide on what they could buy in Australia with the coin. The focus here is currency rates and conversions.

Maps for the commander. A number of years ago, Jan de Lange from the Freudenthal Institute in The Netherlands introduced a historical context, where three spies were sent to draw respective views of a city (surrounded by a circular wall) from the South, West, and North-East, respectively, prior to an army overtaking the city. The drawings from the South and West are provided, but the spy drawing the view from the North-East never returned. The students are asked to draw the view which this spy would have drawn, given the other two drawings. The mathematical focus is views of objects from different positions.

Block of land. Students are presented with a genuine letter to one of the authors of this book, seeking information on the dimensions of a rectangular block of land, given its actual area, and the dimensions of the scale drawing of the block. The mathematical focus relates to scale, areas of rectangles, and maintaining proportions.

Music Cards

Students are asked to analyse which of two music cards provides better value for money—Pod Tunes or New Tunes. This contemporary context engages students in problem solving, with more than one mathematical approach resulting in the correct solution.
Which card is better value?

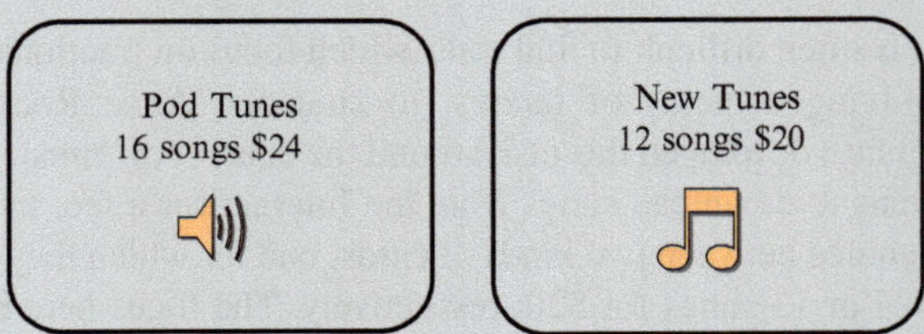

Source

Anne Roche, Australian Catholic University (Melbourne).

Materials

Teachers will need to decide whether they offer students calculators to assist with their calculations, or whether not doing so will be more revealing of their understanding of how to deal with remainders.

Mathematical Focus

Comparative value is one of the most important notions in the use of mathematics in everyday contexts. In this case, students are presented with a situation where they need to be able to compare the relative value for money of two possible purchases. Proportional reasoning is a difficult topic for many students in the middle years, and this task provides the chance for them to see the way in which a variety of calculations (e.g. songs per dollar or dollars per song) can all yield a convincing answer.

Insights from Teaching

Students benefited from group sharing at the end of the activity, as they were able to see the various strategies employed by their peers. Many students and teachers commented that they were surprised at the number of different strategies that could be used to determine the best value music card. Teachers reported that students enjoyed the activity (in particular, the "popular culture" context).

Solution Methods

1. Comparing the cost per song for each card

   Pod = $24 \div 16 = \$1.50$ per song
   New = $20 \div 12 = \$1.67$ per song

2. Comparing how many songs for the same value in cards—e.g.

   Pod = $\$24 \times 5 = \$120$. $16 \times 5 = 80$ songs for \$120.
   New = $\$20 \times 6 = \$120$. $12 \times 6 = 72$ songs for \$120.

Potential Student Difficulties

Some common errors exhibited by students include the following:

- When working out the cost per song, students used the remainder from the division as the decimal value. For example, for Pod, $24 \div 16 = 1$ r 8 but 1 remainder 8 was written as 1.8, and students sometimes concluded that this card (or both cards) were worth \$1.80 per song.
- Identifying the difference in songs or cost between the cards (i.e. four songs difference and \$4 difference), and concluding that this similarity meant that both cards were the same value.

Possible Enabling Prompts

- Focus on just one card (e.g. Pod), and ask what else can we say about the price of the songs from this card? How much would eight songs be? How much would four songs be? How much would two songs be?
- Having solved these problems, ask how this might be helpful for comparing the value of both cards.

Extension Suggestions

- Consider the strategy you chose to calculate the card with better value. Now find at least one alternative approach that will result in the same solution. Write a general process that will enable a student to follow it to solve any problem of this kind.
- Create and solve your own problem involving comparative pricing. Try to make it more difficult than the Music Cards task.

Student Work Samples

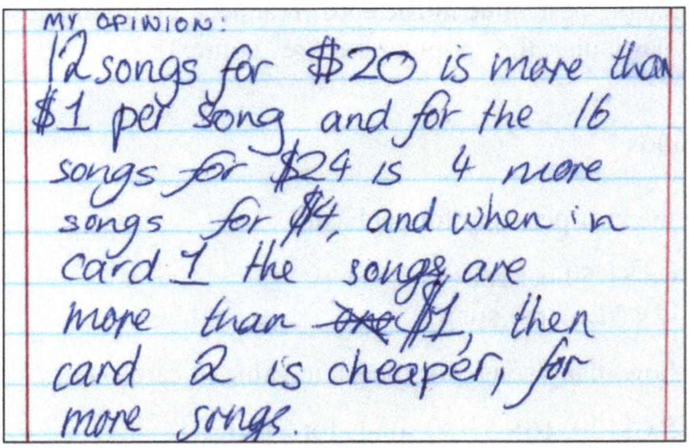
MY OPINION:
12 songs for $20 is more than $1 per song and for the 16 songs for $24 is 4 more songs for $4, and when in card 1 the songs are more than ~~one~~ $1, then card 2 is cheaper, for more songs.

We were impressed with this student method. We believe that what the student is explaining is that both cards involve songs costing at least one dollar each. Therefore if one card provides an additional four songs for an additional four dollars, this must represent improved value, as it must be reducing the average cost of a song.

The Next Lesson

It seems appropriate to initially provide a number of similar comparative value problems, even including more Music Cards examples, to give students more practice at the one context, before moving on. The extension above, involving students creating and solving their own comparative shopping problems, could be adapted, to students in pairs or other small groups posing their problems to other students to solve.

Mike and His Numbers

The newspaper story reports that over more than 2 years, Mike from Devenport wrote all the numbers from one to one million How many digits did he write?	Mike has had the strangest obsession with numbers—but it is out of his system now. For the past 2 years, Mike has hand-written all the digits from one to a million, finishing yesterday. Even Mike doesn't know how to identify his past time—it is hobby, habit, and fascination with numbers rolled into one. It all began two years ago when Mike was watching the Commonwealth Games on TV. "It was late at night, and there were only a few events I was interested in", he said. "So I just picked up a piece of paper and a pen and started writing—one, two, three…" Yesterday afternoon, Mike reached one million. His wife Ruby, who he says has been most tolerant for the past two years, counted down the final ten. Mike says a million will do, and has no other figure-writing landmarks to chase. He has written to the Guinness Book of Records to see if handwriting numbers to a million has been done before. To reach the magic figure has taken 40 96-page exercise books, at a cost of $1 each. Mike wrote the numbers down 20 columns across double pages, and there are 26 lines to a page. He has used 97 ballpoint pens, costing a total of $66.50. If all the figures Mike has written over the past two years were placed side-by-side, it would stretch for 176.36 km. The total number of single digits Mike has written is 5,878,936. It has taken 1,292 hours to write to one million. If he worked on it for eight hours a day, seven days a week, it would have taken 26.3 weeks. Mike says writing the figures has given him an appreciation for how many a million really is. If he were to write to one billion (a thousand million) at the rate it took to write the first million, it would take 2,500 years! His number writing has not been a waste of time. He can tell you what brands of pen last, and which ones blotch. He will now take up a more conventional time filler—jigsaws. He starts his first 3,000-piece puzzle today

Source

The original article about Mike's efforts was in *The Advocate* newspaper from Devonport, in Tasmania, Australia, in 1989. The task was developed by Doug Clarke.

Mathematical Focus

Interestingly, if teachers asked students to imagine writing all the numbers from one to one million, and to calculate the number of digits involved, they may well show little interest. But when they hear about a person in Tasmania actually doing so, the task takes on a new interest. As with many good tasks, there are little elements which prove more difficult than might be expected. In this case, working out how many two-digit whole numbers there are is surprisingly difficult for many students. The mathematical focus includes identifying and using number patterns.

Insights from Teaching

This task is well suited to students who have had experience with problem solving. The task was reported as taking students between 50 min and 2 h to complete, although many students in the shorter sessions could not complete the task.

Solution Methods

The most common solution method took the following form:

Number range	Number of numbers	Number of digits	
1–9	9	9	(1×9)
10–99	90	180	(2×90)
100–999	900	2,700	(3×900)
1,000–9,999	9,000	36,000	(4×9,000)
10,000–99,999	90,000	450,000	(5×90,000)
100,000–999,999	900,000	5,400,000	(6×900,000)
1,000,000	1	7	(7×1)
Total		5,888,896	

Potential Student Difficulties

Many students had difficulty starting without assistance. When faced with the problem, some students found it difficult to see the connection between the task and their understanding of numbers and place value, and had difficulty seeing number patterns. A number of students also devised strategies that were not correct, often because they came to the conclusion that there were 89 numbers from 10 to 99 (inclusive). In order to get the students to reconsider this amount, the teacher might ask the student how many numbers are there from 1 to 99 ("99 of course"). The student may then realise that the 9 single digit numbers, and their suggested 89 double digit numbers only add to 98! As students search for the missing number they will hopefully recalculate that there are 90 two digit numbers between 10 and 99. Another common source of error was from students not adding the last seven digits (from the number 1,000,000).

Possible Enabling Prompts

Ask students:

If Mike wrote all the numbers from 1 to 20, how many digits would that be, and how did you work that out?

How can we work out how many two digit numbers he wrote altogether, and how many digits that would involve?

Extension Suggestions

Mike used 97 ballpoint pens to write out his numbers. On average, how many characters can the ink in one pen write, given a character of text is a single letter or number? Look over your mathematics workbook and estimate how many mathematics workbooks one pen might last you.

Try to develop a formula (in words or symbols) which enables you to determine the number of digits involved in writing all the numbers from 1 to x, for any x.

Student Work Samples

An example of the typical solution was as follows:

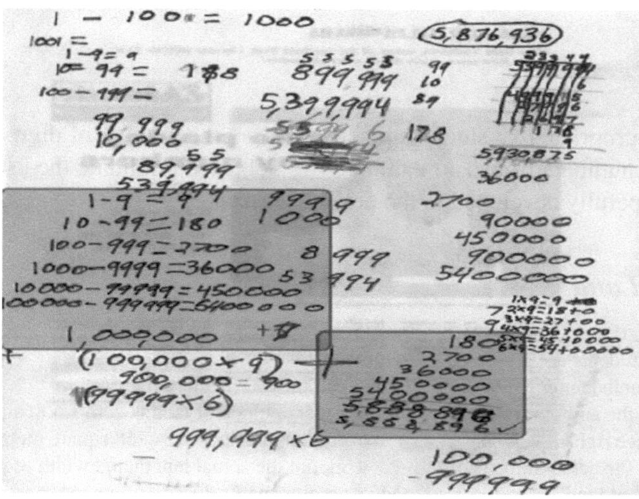

In the next piece of work, the confusion of how many two-digit numbers from 10 and 99 is evident.

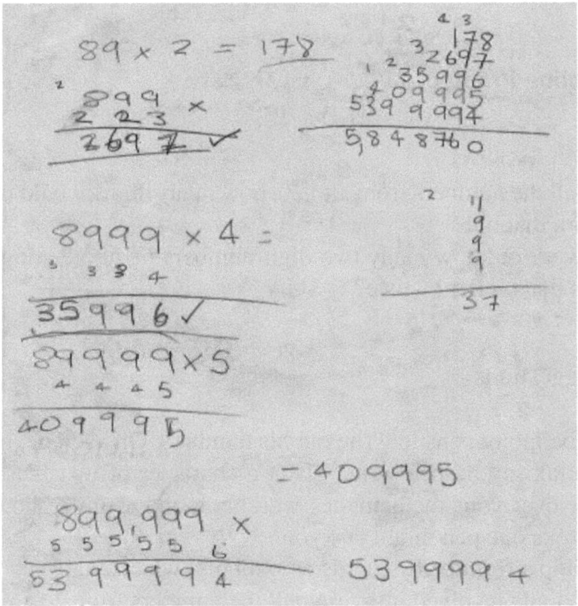

This student has identified a systematic strategy for solving the problem that identifies and uses number patterns, and has chosen multiplication over repeated addition. However, he/she was unable to correctly identify how many numbers are in a range of numbers (for example, between 10 and 99 inclusive = 90 numbers, not 89), therefore not arriving at the correct answer.

The Next Lesson

It seems appropriate that students practise finding the number of digits involved in writing the numbers from 1 to 2,000 or 1 to 550, to consolidate the use of number patterns hopefully developed in the previous lesson.

Block of Land

In this task, students are presented with a genuine letter received by one of the authors from a relative (see right), seeking advice on the actual dimensions of a block of land	Hi Doug, Can I call on your maths expertise? If, on paper, a block of land is 2cm x 5.8cm, and the overall dimensions are 4768 square metres how do I work out the actual length and width of the block? My year nine maths doesn't allow me to work this out! Hope you can help. Thanks Peter Whitten, Editor, RallySport Magazine

Source

A personal email to Doug Clarke, Australian Catholic University (Melbourne).

Mathematical Focus

Scale is one of the most difficult topics for students in the middle years of school. In so many situations faced by students in classrooms, the scale is either given or ignored. Similarly, proportional reasoning, which has been described as the cornerstone of middle years mathematics, provides particular challenges for students, as it draws upon a range of topics which students find difficult in their own right. In this task, students have to preserve the proportions of the scale drawing, 2 cm × 5.8 cm, and determine the dimensions of a rectangle which has an area of 4,768 m^2. Potentially, the task involves area, algebra, proportion, conversion of units, and scale. In the discussion below, most students presume that the block is rectangular, and of course this is not necessarily so. However, at these levels of schooling, other assumptions are probably too difficult to investigate.

Insights from Teaching

Of the five tasks detailed in this section, this was the most difficult for students. It was important in many classrooms for the class to discuss what the problem was actually asking. One teacher posed the following questions to grade 8 students before they started work:

- How might I approach the problem?
- What information do I need to solve the problem?
- What needs clarification before I begin?

Grid paper and calculators seemed helpful materials for students.

Solution Methods

Guessing what the side lengths might be (then multiplying the length by width) and checking if the result is close to 4,768 was common. While some initial guesses in the trial were unrealistic (1,000 × 800!), the process of adjusting the numbers in order to get closer and closer to 4,768 is valuable in allowing students the time to make sense of the problem. During this process, it is important that students consider the proportion of the length to width.

An algebraic solution—students notice that the length is 2.9 times the width and therefore the general solution of width × 2.9 width = 4,768 is true. Therefore,

$$2.9 \times w^2 = 4{,}768$$

$$w^2 = \frac{4{,}768}{2.9}$$

$$w^2 = 1{,}644.14$$

$$w = \sqrt{1{,}644.14}$$

$$w = 40.55 \text{ m}$$

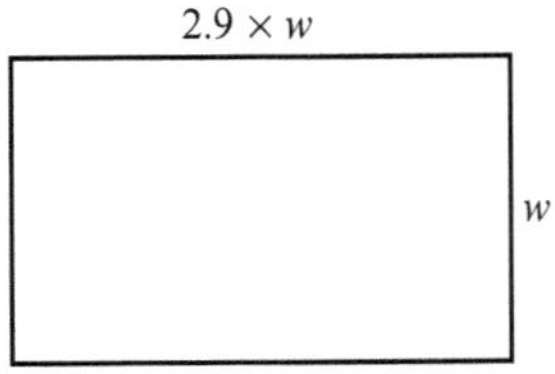

$$l = 2.9 \times w$$

$$l = 2.9 \times 40.55$$

$$l = 117.6$$

Very few students used this method, without extensive prompting and support.

Potential Student Difficulties

Many students had difficulty in determining the length and width of the block. Incorrect approaches observed included the following:

- Dividing the area (4,768 m^2) by 4 giving 1,192 m per side, or 4,768 divided by 2 then 2 again giving them 2,384 m for the long sides and 1,192 m for the shorter sides
- One student noted that the area of the block on paper was 2×5.8 (11.6) but wasn't sure how this might connect to the actual area of the block (4,768 m^2) and,
- One student suggested the sides were $\sqrt{4{,}768}$, giving them 69 m per side and a square block

Possible Enabling Prompts

- Please draw me a rectangle on the grid paper which is 2 cm by 5.8 cm
- Two numbers multiply together and the answer is 4,768. What might the numbers be? (then, using your calculator, try and come up with the dimensions of a rectangle which has an area of 4,768 m^2)

Extension Suggestions

The total area of the state of Colorado in America is 269,618 km^2. Map 1 provides the scale that is connected to this map. Using the scale, calculate the length and width of this state. Compare your two answers and give an explanation if the two results are different.

Map 1: Colorado

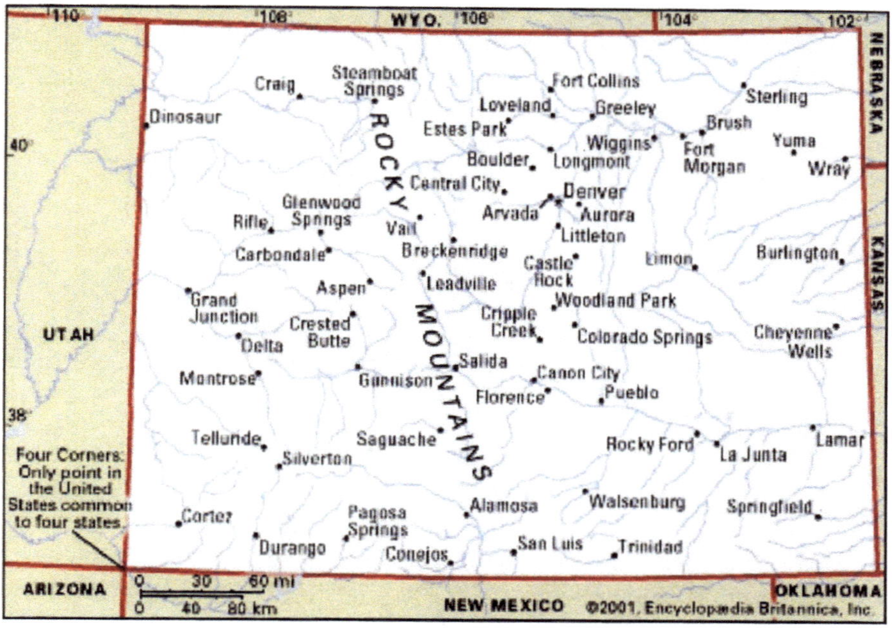

Kakadu National Park is an Aboriginal cultural landscape that has been occupied by Aboriginal people for at least 50,000 years. It is located in the wet-dry tropics of northern Australia, and covers an area of 19,804 km^2.

The following map of Kakadu is a correctly scaled diagram of the national park, and can be downloaded from:

http://www.environment.gov.au/parks/kakadu/images/kakadu-map-large.gif.

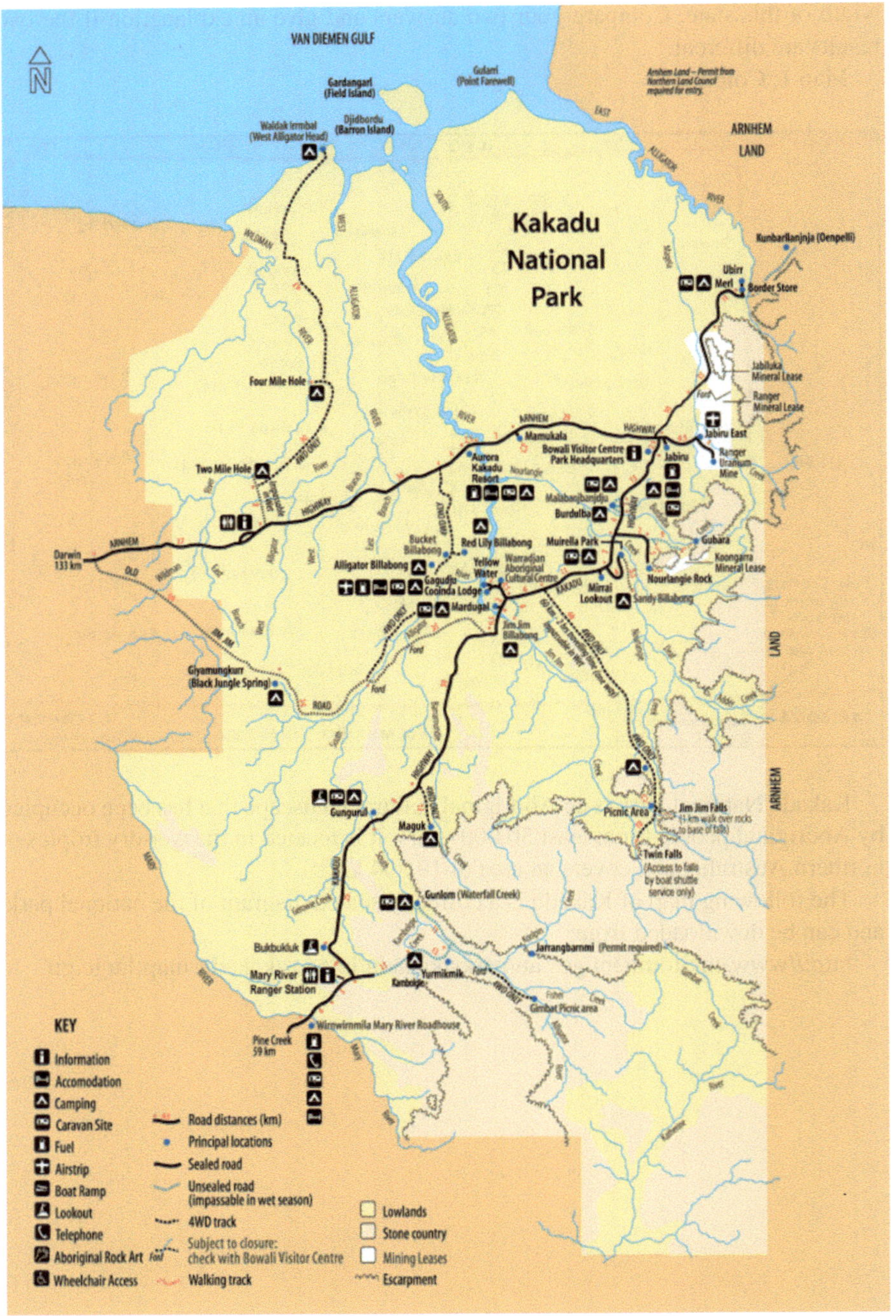

What are the approximate dimensions of Kakadu National Park? Hint: You can split the area up into approximate smaller regular shapes!

The Next Lesson

Given the difficulty of the original task for most students, it may be necessary to consolidate students' understanding of the basic task, by providing a number of practice tasks for students involving creating rectangles where the actual area is given, but the relationship between the sides is also given. This could then be extended to blocks of land, where the blocks are parallelograms, for example.

Comparing Coins from Different Countries

Students are each given a collection of coins from a particular country, with many different currencies being present across the class. Students use online currency converters and newspapers to investigate and compare the value of international currencies in relation to the Australian dollar.

Materials

A range of international coins, Internet access, current newspapers, and calculators.

Source

Doug Clarke, Australian Catholic University (Melbourne).

Mathematical Focus

There is an increasing emphasis in the media on the value of particular currencies in relation to each other, and the steps various governments and reserve banks take to influence these rates. The focus is on understanding how to convert one currency into another.

Insights from Teaching

In most classrooms, teachers sought feedback on students' experiences with foreign currency, e.g. travelling abroad, sending money overseas to relatives, and buying international products online. Because these values are constantly changing in relation to one another, it is necessary to consult the daily exchange rate when calculating how much a foreign currency is worth in Australian dollars or vice versa.

After introductory discussion, one successful way to continue was for the teacher to give each student three coins and ask them to answer the following questions for each coin:

1. Draw your coin and describe it (name, colour, size, what is pictured on it, and what is written on it)
2. Name your coin (e.g. one New Zealand dollar)
3. What is your coin's value in Australian dollars (AUD)?
4. What could you buy in Australia with your coin?

Solution Methods

The students use online currency converter sites like http://www.xe.com to find the value of the coins. Alternatively, they can search the newspaper business section or Google the name of the coin. In small groups, students line up their coins in value compared to the Australian exchange rate. Who has the most/least valuable coin in the class and what is it worth? The students share their discoveries with the class.

Potential Student Difficulties

During the reflection time at the end of the lesson, some students' confusion was revealed about the nature of exchange rates (e.g. if the exchange rate with NZD (New Zealand Dollar) is 1.25, does this mean that $1 Australian dollar is worth NZD 1.25 or NZD 0.80?). The teacher may encourage the students to complete a chart that highlights the exchange rates in a uniform way. Students may be asked to give examples of what 1 AUD and 1 NZD (AUD 0.80) would buy locally to assist in showing that currently a New Zealand dollar is not worth as much as an Australian dollar.

Extension Suggestions

For students who would benefit from additional challenges:

- Make a chart or graph to be updated daily/weekly of the exchange rate of their selected currencies. This might be extended to following the price of shares in gold and oil.
- Students select a fantasy item they might like to buy, e.g. luxury car. Students research the cost of these cars in foreign countries to determine which country would give the best value for money if you paid for the item in Australian dollars. For the budding accountants in the class, refer them to the Luxury Car Tax on the Australian Taxation Office website to establish whether buying a car overseas is worth the money (and effort).

Student Work Samples

This student is examining and describing the features of each foreign coin.

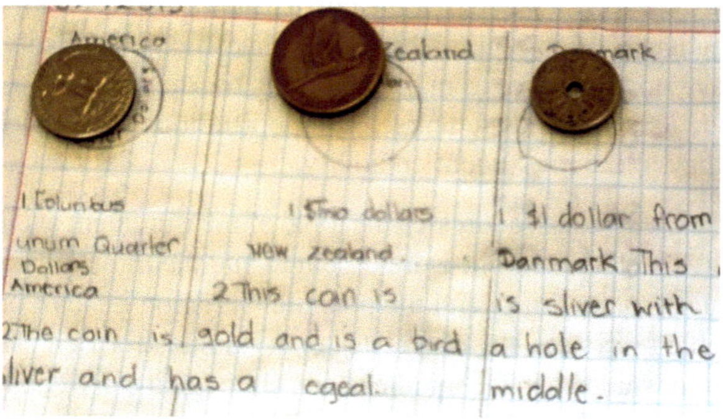

This student used an online currency converter to compare the value of his Hong Kong dollar against the Australian dollar. He explored the purchasing power of the Hong Kong dollar in real-life terms, calculating how many McDonald's ice cream cones it would buy in Australia.

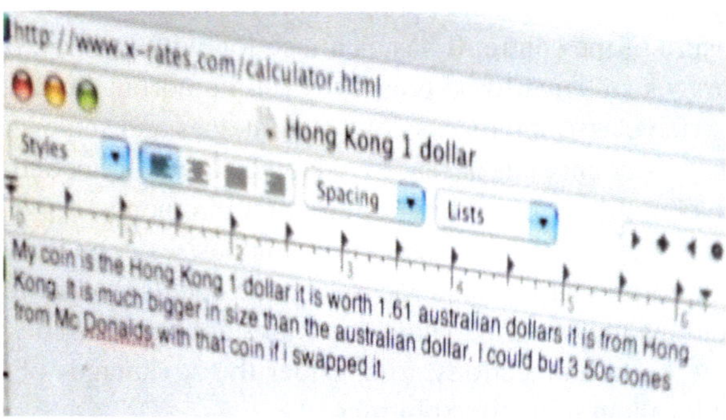

The Next Lesson

This task actually took a number of lessons for students to grasp a sense of how the different currencies related. Follow-up activities could involve students exploring the reciprocal relationship between currencies. For example, if the British pound is worth 2 Australian dollars, how many pounds would be equivalent to 1 Australian dollar?

There could also be a fun competition between groups, where each group is given five different currency coins, and they are timed as they seek to find the total value of the five coins in a sixth currency.

Content-Specific Open-Ended Tasks

This section presents five open-ended tasks that were used by teachers in our classroom trials. These tasks have the usual characteristics of open-ended tasks in that

- There are multiple possible correct responses
- They address particular aspects of the curriculum
- Students engage in important mathematical investigations in solving the tasks
- Students have opportunities to make decisions for themselves
- They provide teachers and students with opportunities for mathematical discussions, and they allow for student creativity

The *Wrap the present* task addresses the topic of measurement and in particular the relationship between the linear measure of length and the three dimensional box. We have used this task many times, and the potential variation in solution strategies and the many possible responses always creates opportunities for rich classroom discussions. It is ideal for students in the middle years, although it can be adapted to students at higher levels by using a more complex shaped present, such as a triangular prism.

The *Painting a Room* task is about surface area of a three-dimensional shape. It has a practical and relevant dimension making it similar to contextualised tasks, but the opportunity for students to make their own decisions connects it to this type of task. As it is written, it is suitable for middle years students but could easily be adapted for later levels by increasing the complexity of the information provided.

The *Writing a Sentence* task encourages students to consider the meaning of the mean as a measure of the centre. It has a connection with literacy, but more importantly it encourages students to be creative, even humorous, and allows plenty of opportunity for rich classroom discussions of strategies and sentences. It is suitable for upper primary students but is readily adapted to junior secondary students by making it more complex or by adding different requirements.

The *Different Ways to Represent Data* task emphasises connections between different graphical representations, and focuses on the details that are required to make such representations meaningful. There are important opportunities for students to make choices, to build connections, to consider the advantages of different representations, and for them to justify solutions.

The *Money Measurement* task focuses on measurement and has slightly different characteristics from the others. It is open ended in that there is a range of ways students can tackle the task and a range of levels of accuracy that they can seek. It also has characteristics of a contextualised task. In fact, we found that many tasks can be categorised in multiple ways. This does not detract from the classroom potential of such tasks. It can be challenging over a range of levels from the year 3 to year 12 depending on the specificity of the strategies expected from the students. Overall, the five tasks and the associated descriptions offer further insights into the ways such tasks can be used in classrooms.

Wrap the Present

You need to wrap a present in a box. You have 1 metre of ribbon. The bow at the top will use 30 cm of this.
What might be the dimensions of the box?

The students are required to select relevant information from what is given, solve the problem, see that multiple solutions are possible, and describe a range of solutions

Source

Peter Sullivan.

Mathematical Focus

The task assumes that students know units of length, can measure length, and that they know the meaning of the word "dimensions".

The task encourages students to connect their knowledge of length and units to a 3D object, connecting 2D representations to the 3D shape, and visualising the rectangular prism.

Insights from Teaching

It is helpful for students to work in small groups if they are struggling with this task. One of the highlights of this activity in one of the trials was the sharing of students' strategies as they were created. This helped and encouraged students who were having difficulty progressing with the task.

Teachers who were involved in trials, considered the task to be an excellent indicator of each student's problem-solving approaches.

Solution Methods

The following is an example of a common solution method. Note that most students quickly recognised that the bow could be excluded from the total:

There are 4 box heights, 2 lengths, and 2 widths for the string. Assuming the height is 5 cm (there are 4 of them) and the length is 10 cm (there are 2 of them), that leaves 30 cm (out of the 70 cm left), that means that the width is 15 cm. These can be progressively varied for other solutions.

Some students used algebraic methods ($4H+2L+2W=70$), and some students created tables of values.

Potential Student Difficulties

In the trial, a number of students had difficulty commencing the task without additional support. Students who had difficulty commencing became disengaged quickly, as they felt the mathematics was beyond them. Some students had trouble identifying the relevant information from the instructions, e.g. 1-m ribbon and a 30-cm bow.

None of the students in the trial considered a box that was not a rectangular prism, and many arrived at similar answers.

It is also worth emphasising that the diagram is not drawn to scale, and they should not use the diagram as a guide.

Possible Enabling Prompts

The following suggestions can be made to students who are experiencing difficulty:

- Here is a box (a real one) wrapped and tied with string. Work out how long is the string without untying it.
- Assume your box is a cube. How many faces will the ribbon need to pass over (you may want to find a real box to count the number of faces)? How much ribbon is available to wrap these faces? What are the dimensions of this box?
- Assume for now that the height is 5 cm.

Extension Suggestions

The following suggestions can be made to students who finish quickly:

- Describe as many solutions as possible.
- Now you have a box that takes 2 m of ribbon. What might be the dimensions of this box?
- What if the box was a triangular prism? What would be the possible dimensions? How would the string be tied?

Student Work Samples

This student, as did many others, only considered one example for this problem. Extra time was needed to be allocated to the task.

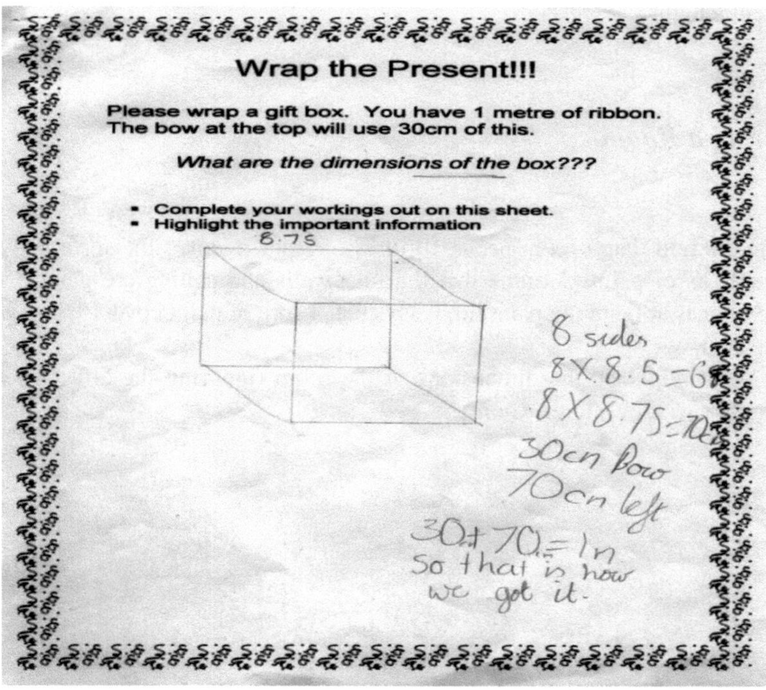

The student below attempted a generalised solution but has not given specific possible solutions.

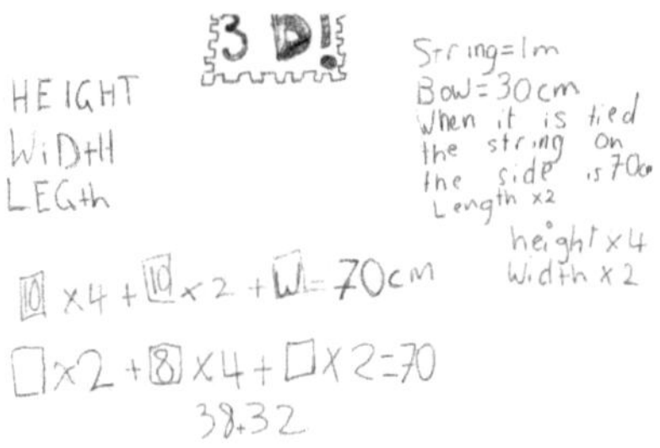

The Next Lesson

One possibility for the next lesson could be to extend the modes of recording their approaches including the drawing of diagrams such as thinking about the net of the box. Another possibility could be to use boxes of different shapes and/or string of different lengths.

Painting a Room

You are told that a room needs 10 litres of paint to cover the surfaces with two coats of paint. Assume that the four walls and ceiling are painted but the floor is not. On the paint tin, it says that 1 litre of paint covers 16 m^2 with one coat.

What might be the dimensions of the room (ignoring the effect of any doors or windows in the room)?

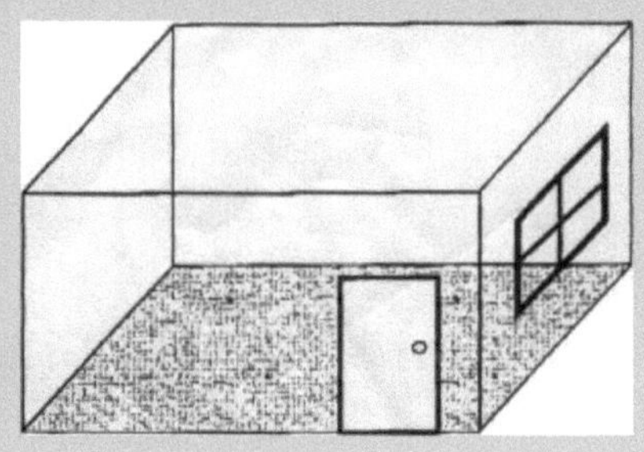

Source

Project teachers.

Mathematical Focus

The focus of the problem is to see that the area of a shape, and in this case surface area, does not determine its dimensions, and that for a given area various dimensions are possible.

The students also explore ways of visualising or drawing this three-dimensional shape in two dimensions, as well as recognising the properties of prisms that help to define the problem, such as that opposite walls of a rectangular room are the same size and shape.

There is a substantial problem-solving demand, in that students need to process various assumptions and pieces of information, in determining their solutions.

Another important focus for students' learning is the way they will record and communicate their solutions.

Insights from Teaching

The teachers who used this task found that it is well suited to the conclusion of a unit of work that encompasses surface area.

Teachers in the trial reported that it would be helpful to pose similar, but easier, independent problem-solving activities as a precursor to this task, to assist students in tackling the multi-step nature of activity.

Solution Methods

The first step is to realise that the total area that can be covered by the paint is 160 m^2 so each coat will cover 80 m^2

After that possible solutions to this activity are (from actual student work):

Assume that the length, width and height are equal. There are 5 squares, each square is 16 m^2, therefore length, width and height are all 4 m

Another approach is to use trial and error.

Assume length = 4 m, width = 5 m, height = 3 m. Area = $1\times20+2\times12+2\times15=74$ m^2, which is too small

Assume length = 4 m, width = 5.5 m, height = 3 m. Area = $1\times22+2\times12+2\times16.5=79$ m^2— which is closer

An algebraic approach might be:

Make the length l, width, w and height h

The surface area is $2lh+2wh+lw$ which equals 80 m^2

Assume the height is 3 m, then

$6l+6w+lw=80$

Now assume that $l=5$ m

$30+6w+5w=80$

$11w=50$

So $w=4.54$ m

and so on.

Potential Student Difficulties

In the trial, some students found it difficult to complete the task through to presentation of an answer. A number of students were unable to provide an explanation of their mathematical thinking, as independent problem solving of this complexity was a relatively unfamiliar experience for them.

Teachers observed that this task was challenging for some students. However, those students who proved unable to logically think through any process to provide answers enjoyed the lesson conclusion when students who did come up with answers presented their logic and conclusions to the class group.

All students in the trial were reported as being engaged and committed to solving the task. The students were also excited and interested in the outcomes of their peers.

Possible Enabling Prompts

Some suggested prompts to assist students experiencing difficulty are as follows:

- For a start, assume that the standard height of a room is 3 m.
- You have been given 10 litres of paint to paint one rectangular wall. What might be the dimensions of the wall?
- You have been given enough paint to paint a rectangular wall of area 36 m^2. What could be the dimensions of the wall?

Extension Suggestions

Some possible prompts to extend students who have finished include the following:

- Now work out some possible dimensions if the room has one door of area 2 m^2; and one window of 3 m^2.
- What solutions can you find that use only dimensions in metres (that is with no centimetres)?
- What is the largest room that you can paint?
- What would the smallest dimensions that the room might reasonably have?

The Next Lesson

It would be useful to extend this activity to other aspects of surface area, such as calculating the surface area of cubes and rectangular prisms of given dimensions. The reverse is also possible. The students can be given the surface area of a rectangular prism and asked to determine the dimensions.

Writing a Sentence

A person wrote a sentence which contained only five words.
The average number of letters in each word was 4.
However, none of the words has four letters.
What might the sentence have been?

Source

Sullivan, P., & Lilburn, P. (2004). *Open-ended maths activities: using 'good' questions to enhance learning in mathematics* (2nd ed.). South Melbourne: Oxford University Press.

Mathematical Focus

The task assumes that students know the meaning of the word "average". If not, this may require clarification before commencing the task.

The implication in the task is that students will create their own sentences. The task also encourages students to devise their own ways of approaching the task. It also allows students to gain insights into the two ways of thinking about average: one method (in this case) is that the average is the sum of the letters divided by the number of words; the other method is that the individual scores are evenly spread above and below the average.

Insights from Teaching

In the trial, the teacher encouraged the students to find more than one solution to the problem. Students were given time to work on the task. After 5–10 minutes, the teacher stopped the students and asked three or four students to share their strategies with the class. Some students mentioned the total of 20 letters.

The teacher noted that the students responded successfully with a range of possible answers to the question, demonstrating their understanding of the term *average*.

Solution Methods

Students created sentences like the following:

Red lobsters nip old men.
Bubbles float on hot air.
A sad owl flies silently.
Jam is an acquired taste.

The most common approach was from those students who realised that the total number of letters would be 20, and created five words with the total of the letters being 20.

There were students who wrote down five 4-letter words and then progressively increased the number of letters in some words, and reduced the others by the same amount.

Potential Student Difficulties

The teacher recognised that some students had difficulty with getting started on this task as they were unsure how to approach the problem. While many students had success with the problem once they had realised the significance of the number 20 in the task, others required further assistance.

Possible Enabling Prompts

The following prompts can be made to students who are experiencing difficulty:

- Ask them to create a sentence with just three words.
- Ask them to write any sentences using 10 letters, 15 letters.
- Write any sentence and work out the total of the letters.
- Write down five numbers that have an average of 4.

Extension Suggestions

The following suggestions can be made to students who finish quickly:

- Use a book to find sentences that fit the criteria of the question.
- Place constraints on the task, such as you must have at least one 6-letter word, or two 3-letter words, etc.

- Can you make a new solution by adapting one of your existing solutions?
- Find a way to describe all possible solutions.

Student Work Sample

The following is a sample of a student response. This student has demonstrated an understanding of the term *mean*. S/he has first calculated the total number of letters in the sentence (20) and is manipulating the numbers in each word to fit the criteria. However, the student will need to review the responses as some include four-letter words. It appears that at this stage of the task, the suggestion of not including four-letter words has been overlooked.

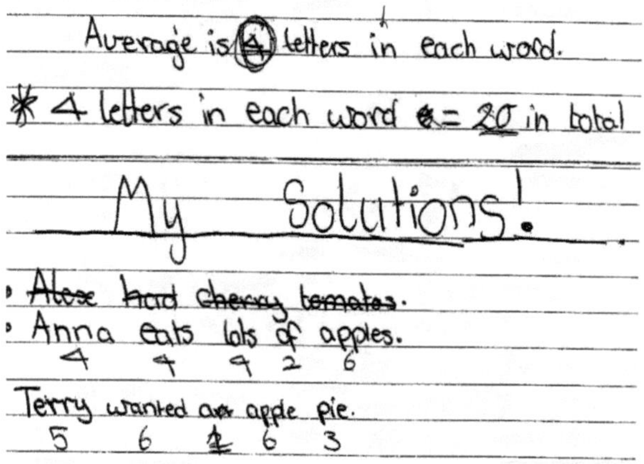

The Next Lesson

One possibility for the next lesson is for the students to do a similar task in a different context, such as “The average number of points per game of a group of five basketball players was 10. What might be the number of points they scored?” Another possibility is to ask the same question as in this lesson but using median or mode (or both).

Different Ways to Represent Data

Students are given a pie graph which represents the results of a survey of students' favourite sports played at a particular school. Students are asked to represent these data in different ways.
This is an open-ended task where students choose their own method of presenting the data and are encouraged to pursue multiple representations.

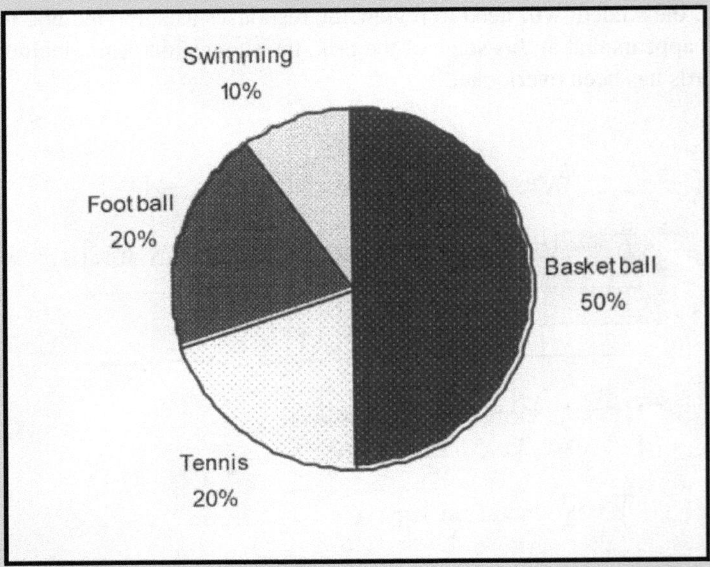

Source

One of the lesson sequences from the project teachers.

Mathematical Focus

The task assumes that the students can interpret the pie chart, can recognise the pieces as proportional, and can relate the percentages given to the total sample in the survey.

The task prompts students to recognise that information presented in some graphs and tables provides the total number of responses. In this case, students have to make assumptions about the number of responses for themselves.

The task prompts students to think about alternate ways of presenting data and to see connections between those different representations.

Some of the mathematical processes on which students can draw are using decimals, ratios, and percentages to find equivalent representations of common fractions (for example, 3/4 = 9/12 = 0.75 = 75% = 3:4 = 6:8), using different graphical forms, and distinguishing between categorical and numerical data.

Insights from Teaching

Teachers who used this task reported that it is well suited as an introduction to a unit on data, to assess prior knowledge of the variety of data presentation forms (e.g. line graphs, bar graphs, and picture graphs).

The teachers suggested that some students may benefit from revision of conversion between numbers, percentage, and fractions prior to undertaking this task.

Teachers found that it was helpful to remind students to label all diagrams, and found the task to be useful in identifying areas for future teaching.

Most students reported this task to be relatively easy, and they identified the teaching outcomes as being related to learning about different graphical presentations.

Teachers observed that student engagement was high, and the problem of transfer from one presentation type to another encouraged their continued application to the task. They also found that most students enjoyed the autonomy of being empowered to demonstrate their knowledge.

Some teachers reported that students used Excel to draw their graphs.

Solution Methods

As a result of the open-ended nature of this task, there are a number of possible correct responses to this problem. The work sample below is a suitable illustration of one such response.

The challenge is to encourage students to respond with a variety of types of graphs.

Potential Student Difficulties

Teachers reported that some students in the trial showed discrete data in a continuous graph; that some students needed additional support when converting between percentages, fractions, and numbers; and that students demonstrated a limited choice of graphical representations—the most common being bar graphs and line graphs.

Possible Enabling Prompts

Some suggested prompts to assist students experiencing difficulty in starting the activity are as follows:

Which one of the following column graphs might match the pie graph that you were given?
How did you work it out?
Asking simpler questions such as "Show these data another way"

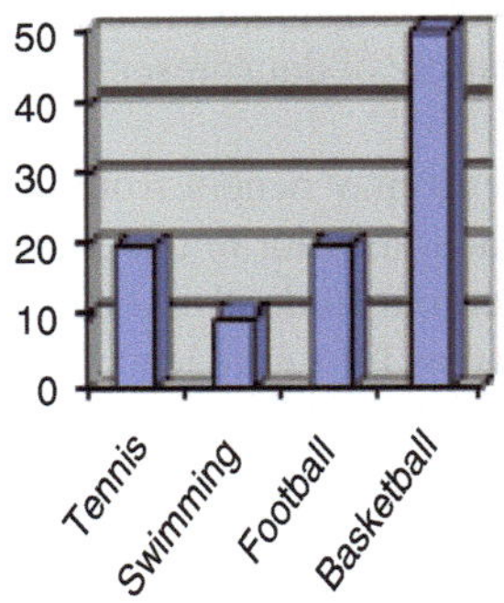

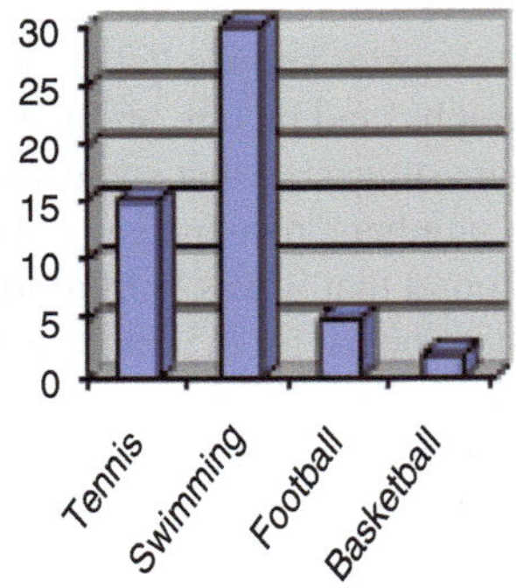

Extension Suggestions

Suggestions for students who would benefit from additional challenges are as follows:

- If there are 85 students who prefer basketball at the school, draw a graph (that is a different type to those already used in this activity) to show how many students prefer each of the different sports.
- What do you think the following graph could be telling us? Identify as much information as possible. Is there anything missing?

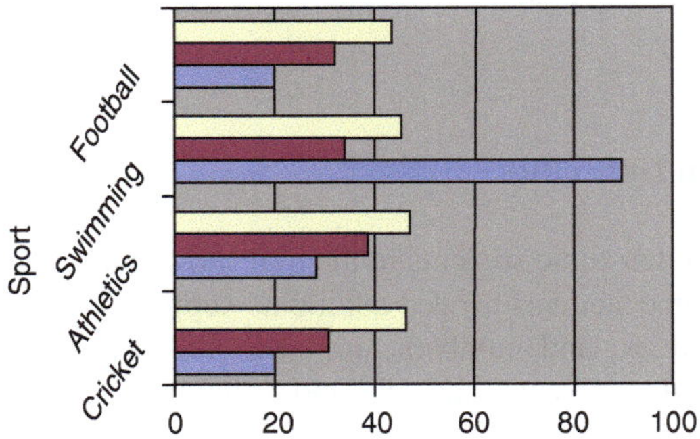

- Convert the data from this graph into a table.
- List all the types of graphs that you can think of, and for each one, identify if you think it is appropriate to use for this information and why.

Student Work Samples

These work samples illustrate different ways that students might draw an alternate representation of the data in the pie chart which was the initial prompt. In each case, they have made an assumption about the number of students being surveyed.

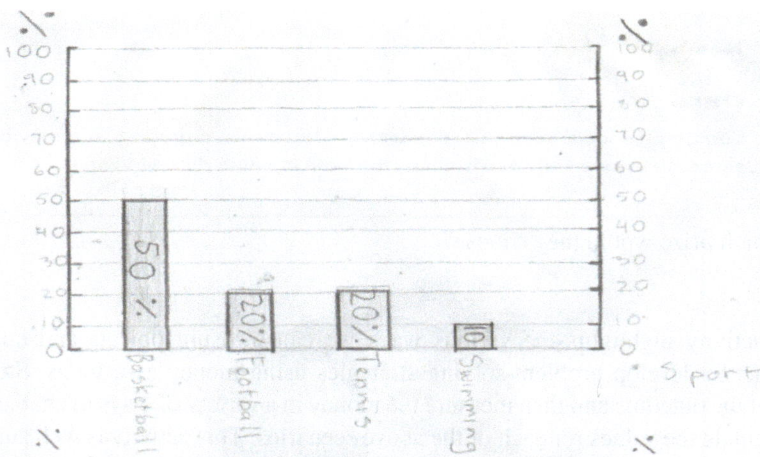

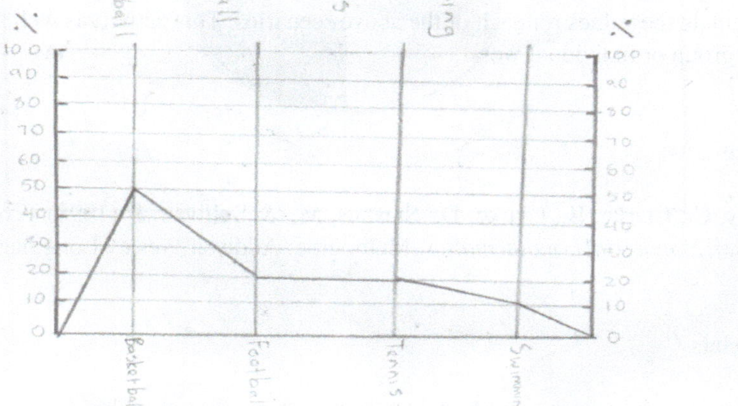

The Next Lesson

Possible ways of extending this experience might be by:

- Having a different graph type as the initial prompt
- Having a more complex graph of the same type
- Being more specific about requiring students to compare and contrast the types of representation

Money Measurement

Students are told they have won a prize. The prize can be one of:

- One metre of $2 coins (lying flat)
- One square metre of 5-cent pieces (edges touching, lying flat)
- A 1-litre milk carton full of 20-cent pieces or
- One kilogram of $1 coins

Which prize would they choose?

This activity highlights the various ways we can measure objects and compels students to develop problem-solving strategies using money as a focus. Students predict the outcome and then measure the money in a variety of ways to enable them to calculate the values for each of the above scenarios. This activity is well suited to either group or individual work.

Source

Beesey, C., Clarke, B., Clarke, D., Stevens, M., & Sullivan, P. (1998). *Exemplary assessment materials: mathematics*. Melbourne: Addison Wesley Longman.

Materials

Metre rulers and tape measures, 1-litre milk cartons, digital scales, balance scales, and assorted coins ($1, $2, 20 cents, and 5 cents).

Mathematical Focus

This task introduces students to various aspects of measurement, although it is only marginally about money. In each of the comparisons and calculations, students have a range of options, many of which focus student attention onto threats to accurate measurements. Students will also see that the answer that they think is most likely is not necessarily the optimal one.

Insights from Teaching

The time taken to complete this activity varied greatly and directly depended on the skill level of students undertaking it in the trial. On average, it took two periods to complete for upper primary/lower secondary aged students.

To reduce the amount of time the exercise takes, it can be run in groups. Students can be grouped together so that each group can solve a different problem. The class needs to regroup to discuss the strategies used.

This activity needs to be well prepared in advance, as it can be challenging collecting the variety of resources needed.

Students will need an opportunity to discuss questions prior to starting the activity. In the trial, many questions such as "how are the coins stacked?" and "will the coins be melted down?" were asked by students prior to being able to commence the activity.

Potential Student Difficulties

Students without sufficient background in measurement of length, area, volume, and mass found this activity challenging, and some students benefited from revision of calculating these measurements. However, many students completed the activity quite quickly, highlighting the necessity for extension activities to be prepared ahead of time.

Common problems observed in the student work samples included difficulty in accurately converting metric units of length, difficulty in starting the task for some of the measurements (e.g. volume or area), and in many cases, difficulty in logically writing down their method (which was highly correlated with erroneous answers). The most common calculation error was in volume, where many of the students approaching the task mathematically made erroneous assumptions (e.g. that the weight of 1 litre of coins is 1,000 g).

Possible Enabling Prompts

- If I give you fifty 5 cent pieces, how much money do you have? If I give you fifty 20 cent pieces, how much money do you have? If I give you fifty $1 pieces, how much money do you have?
- Draw a 10 cm × 10 cm square (100 cm^2). Using a 5-cent coin as a stencil, how many 5-cent circles can you draw with edges touching in this area? Can you think of a way to work this out without counting each coin?

- What are the dimensions (measurements of each side) of a 1-litre container (use a ruler or tape measure)? Choose one side and work out how many 20-cent coins you can fit on this side with the least number of gaps between them. How might this information help you to work out how many coins will fill your 1-litre container?

Extension Suggestions

- How many different three-dimensional shapes can you calculate that will have a volume equivalent to 1 litre? Will the same dollar value of coins fit into each shape? If not, why not and which of your shapes will have the most money in it? Which shape will have the least money in it?
- Which would you prefer—your height in $1 coins or your weight in 5-cent pieces?

Student Work Samples

This student accurately converted between different metric units and clearly articulated the method used.

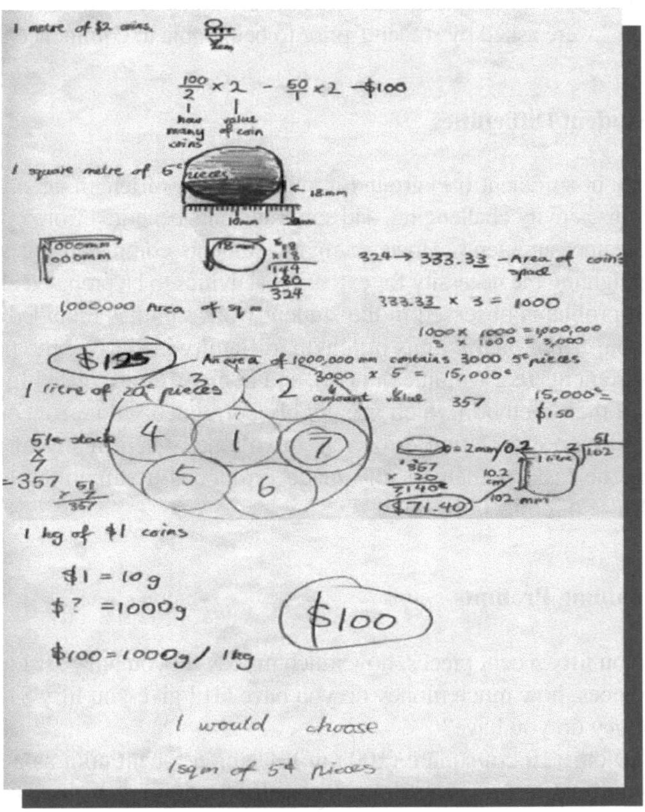

This student has correctly identified a mathematically sound method for working out each of the four challenges:

1 m of \$2 coins by measuring the coin, and determining the number of coins that would fit into the distance and what value that would have. (\$100)
1 m^2 of 5 cent coins by measuring the area of the "coin space" (the square containing the coin) and determining how many would fit into 1 m^2 (and what value that would have). (\$150)—the \$125 remains a mystery.
1 litre of 20 cent coins by physically fitting 7 coins into the base of a 1 litre cylindrical jug (the 1 L milk carton was not specified in the original prompt for this student). The student measured the depth of the coin and determined how many coins would then fit in the jug, and what value that would have in total. (\$71.40)
1 kg of \$1 coins by measuring the weight of a \$1 coin and determining how many would fit into 1 kg. (\$100)

The Next Lesson

It would be possible to do such comparisons with a range of other everyday objects, such as drink containers (how many 1-litre containers of milk would fill up a cubic metre box? What if the milk was tipped into a cubic metre?)

References

ACARA. (2011). The shape of the Australian Curriculum: Mathematics. Found July 1st at http://www.acara.edu.au/phase_1_-_the_australian_curriculum.html.

Allen, B. (2003). Pupils' perspectives on learning mathematics. In B. Allen & S. Johnston-Wilder (Eds.), *Mathematics education: Exploring the culture of learning* (pp. 233–241). London: Routledge Falmer.

Anderson, J. (2003). Teachers' choice of tasks: A window into beliefs about the role of problem solving in learning mathematics. In L. Bragg, C. Campbell, G. Herbert, & J. Mousley (Eds.), *Mathematics educations research: Innovation, networking, opportunity* (Proceedings of the 27th annual conference of the Mathematics Education Research Group of Australasia, pp. 72–80). Geelong: MERGA.

Anthony, G., & Walshaw, M. (2009). *Effective pedagogy in mathematics*. Educational series 19. Brussels: International Academy of Education; Geneva: International Bureau of Education.

Association of Teachers of Mathematics (ATM). (1988). *Reflections on teacher intervention*. Derby: ATM.

Bakker, A., Hoyles, C., Kent, P., & Noss, R. (2006). Improving work processes by making the invisible visible. *Journal of Education and Work, 19*(4), 343–361.

Ball, D. (1992). Magical hopes: Manipulatives and the reform of mathematics education. *American Educator, 16*(2), 46–47.

Bishop, A. J. (2001). What values do you teach when you teach mathematics? In P. Gates (Ed.), *Issues in mathematics teaching* (pp. 93–104). London: Routledge Falmer.

Boaler, J. (1993). The role of contexts in the mathematics classroom: Do they make mathematics more "real"? *For the Learning of Mathematics, 13*(2), 12–17.

Boaler, J. (2002). *Experiencing school mathematics: Traditional and reform approaches to teaching and their impact on student learning*. Mahwah, NJ: Lawrence Erlbaum.

Boaler, J. (2008). Promoting 'relational equity' and high mathematics achievement through an innovative mixed ability approach. *British Educational Research Journal, 34*(2), 167–194.

Boulton-Lewis, G., & Halford, G. (1992). The processing loads of young children's and teachers' representations of place value and implications for teaching. *Mathematics Education Research Journal, 4*(1), 1–23.

Bransford, J. B., Brown, A. L., & Cocking, R. R. (Eds.). (1999). *How people learn: Brain, mind, experience, and school*. London: Committee on Developments in the Science of Learning, National Research Council.

Broekman, H., Van der Valk, T., & Wijers, M. (2000, August). *Teacher knowledge needed to teach ratio and proportion in secondary school maths: On using the ratio table*. Paper presented at the 25th Annual Conference of the Assocation of Teachers Educators in Europe (ATEE), Barcelona.

P. Sullivan et al., *Teaching with Tasks for Effective Mathematics Learning*, Mathematics Teacher Education 9, DOI 10.1007/978-1-4614-4681-1,

Brophy, J. E. (1991). Conclusion to advances in research on teaching. In J. Brophy (Ed.), *Advances in research on teaching: Vol. 2 Teachers' subject matter knowledge and classroom instruction* (Vol. II: Teachers' knowledge of subject matter as it relates to teaching practice, pp. 347–362). Greenwich, CT: JAI.

Brown, S. I., & Walter, M. I. (2004). *The art of problem posing* (3rd ed.). New York: Routledge.

Brown, R., & Renshaw, P. (2006). Transforming practice: Using collective argumentation to bring about change in a year 7 mathematics classroom. In P. Grootenboer, R. Zevenbergen, & M. Chinnapan (Eds.), *Identities, cultures and learning spaces* (Proceedings of the 29th conference of the Mathematics Education Research Group of Australasia, pp. 99–107). Canberra: MERGA.

Butler, F., Miller, S. P., Crehan, K., Babbitt, B., & Pierce, T. (2003). Fraction instruction for students with mathematics disabilities: Comparing two teaching sequences. *Learning Disabilities Research and Practice, 18*(2), 99–111.

Carpenter, T. P., Lindquist, M. M., Matthews, W., & Silver, E. A. (1983). Results of the third NAEP mathematics assessment: Secondary school. *Mathematics Teacher, 76*, 652–659.

Cass, M., Cates, D., & Smith, M. (2003). Effects of manipulative instruction on solving area and perimeter problems by students with learning disabilities. *Learning Disabilities Research and Practice, 18*(2), 112–120.

Charalambous, C. Y. (2008). Mathematical knowledge for teaching and the unfolding of tasks in mathematics lessons: Integrating two lines of research. In O. Figuras, J. L. Cortina, S. Alatorre, T. Rojano, & A. Sepulveda (Eds.), *Proceedings of the 32nd annual conference of the International Group for the Psychology of Mathematics Education* (Vol. 2, pp. 281–288). Morelia: PME.

Christiansen, B., & Walther, G. (1986). Task and activity. In B. Christiansen, A. G. Howson, & M. Otte (Eds.), *Perspectives on mathematics education* (pp. 243–307). Dordrecht: Reidel.

Clark, C. M., & Peterson, P. L. (1986). Teachers' thought processes. In M. C. Wittrock (Ed.), *Handbook of research on teaching* (pp. 255–296). New York: Macmillan.

Clarke, D. M. (2006). Fractions as division: The forgotten notion? *Australian Primary Mathematics Classroom, 11*(3), 4–10.

Clarke, D., Cheeseman, J., Gervasoni, A., Gronn, D., Horne, M., McDonough, A., Montgomery, P., Roche, A., Sullivan, P., Clarke, B., & Rowley, G. (2002). *Early Numeracy Research Project (ENRP): Final report.* Retrieved September 11, 2002, from http://www.sofweb.vic.edu.au/eys/num/enrp.htm.

Clarke, D. M., & Clarke, B. A. (2004). Mathematics teaching in Grades K-2: Painting a picture of challenging, supportive, and effective classrooms. In R. N. Rubenstein & G. W. Bright (Eds.), *Perspectives on the teaching of mathematics* (66th Yearbook of the National Council of Teachers of Mathematics, pp. 67–81). Reston, VA: NCTM.

Clarke, D. M., & Roche, A. (2009). Using mathematical tasks built around "real" contexts: Opportunities and challenges for teachers and students. *Australian Primary Mathematics Classroom, 14*(2), 24–31.

Clarke, D. M., & Roche, A. (2010). The power of a single game to address a range of important ideas in fraction learning. *Australian Primary Mathematics Classroom, 15*(3), 18–23.

Clarke, D. M., Roche, A., & Mitchell, A. (2008). Ten practical, research-based tips for making fractions come alive (and make sense) in the middle years. *Mathematics Teaching in the Middle School, 13*(7), 373–380.

Cobb, P., & McClain, K. (1999). Supporting teachers' learning in social and institutional contexts. In Fou-Lai Lin (Ed.), *Proceedings of the 1999 International Conference on Mathematics Teacher Education* (pp. 7–77). Taipei: National Taiwan Normal University

de Lange, J. (1996). Using and applying mathematics in education. In A. J. Bishop, K. Clements, C. Keitel, J. Kilpatrick, & C. Laborde (Eds.), *International handbook of mathematics education* (pp. 49–97). Boston, MA: Kluwer.

Delpit, L. (1988). The silenced dialogue: Power and pedagogy in educating other people's children. *Harvard Educational Review, 58*(3), 280–298.

Department of Education and Early Childhood Development. (2006). *Maths developmental continuum P-10*. Downloaded from http://www.education.vic.gov.au/studentlearning/teachingresources/maths/mathscontinuum/default.htm.

Desforges, C., & Cockburn, A. (1987). *Understanding the mathematics teacher: A study of practice in first schools*. London: Falmer.

Dole, S. (2008). Ratio table to promote proportional reasoning in the primary classroom. *Australian Primary Mathematics Classroom, 13*(2), 18–22.

Doyle, W. (1986). Classroom organisation and management. In M. C. Wittrock (Ed.), *Handbook of research on teaching* (pp. 392–431). New York: Macmillan.

Dweck, C. S. (2000). *Self theories: Their role in motivation, personality, and development*. Philadelphia, PA: Psychology Press.

Education Queensland. (2010). Productive pedagogies. Downloaded in January 2010 from http://education.qld.gov.au/corporate/newbasics/html/pedagogies/pedagog.html.

Elliot, A. J. (1999). Approach and avoidance motivation and achievement goals. *Educational Psychologist, 34*(3), 169–189.

English, L. D. (1998). Children's problem posing within formal and informal contexts. *Journal for Research in Mathematics Education, 29*(1), 83–106

Erickson, T. (1989). *Get it together: Maths problems for groups: Grades 4-12*. San Francisco, CA: Lawrence Hall of Science.

Ernest, P. (1988). The attitudes and practices of student teachers of primary school mathematics. In A. Borbas (Ed.), *Proceedings of the Annual Conference of Psychology of Mathematics Education* (Vol. 2, pp. 288–295). Beszprem, Hungary: PME.

Ernest, P. (1994). Varieties of constructivism: Their metaphors, epistemologies and pedagogical implications. *Hiroshima Journal of Mathematics Education, 2*, 1–14.

Ernest, P. (2010). The social outcomes of learning mathematics: Standard, unintended or visionary? In C. Glascodine & K. A. Hoad (Eds.), *Make it count: What research tells us about effective mathematics teaching and learning* (pp. 21–26). Camberwell: ACER.

Fennema, E., & Franke, M. (1992). Teachers' knowledge and its impact. In D.A. Grouws (Ed.), *Handbook of research on mathematics teaching and learning* (pp. 147–164). New York: Macmillan Publishing.

Fennema, E., Franke, M. L., Carpenter, T. P., & Carey, D. A. (1993). Using children's mathematical knowledge in instruction. *American Educational Research Journal, 30*(3), 555–583.

Fredericks, J. A., Blumfield, P. C., & Paris, A. H. (2004). School engagement: Potential of the concept, state of the evidence. *Review of Educational Research, 74*(1), 59–110.

Gee, J. P. (2004). *Learning by design: Games as learning machines*. http://labweb.education.wisc.edu/room130/jim.htm. Accessed 25 Apr 2005.

Gravemeijer, K. (1997). Instructional design for reform in mathematics education. In M. Beishuizen, K. Gravemeijer, & E. Van Lieshout (Eds.), *The role of contexts and models in the development of mathematical strategies and procedures* (pp. 13–34). Utrecht: Technipress.

Greer, B. (1997). Modeling reality in the mathematics classrooms: The case of word problems. *Learning and Instruction, 7*(4), 293–307.

Griffin, S., & Case, R. (1997). Re-thinking the primary school with curriculum: An approach based on cognitive science. *Issues in Education, 3*(1), 1–49.

Hannula, M. (2004). *Affect in mathematical thinking and learning*. Turku: Turun Yliopisto.

Hattie, J., & Timperley, H. (2007). The power of feedback. *Review of Educational Research, 77*(1), 81–112.

Hiebert, J., & Wearne, D. (1997). Instructional tasks, classroom discourse and student learning in second grade arithmetic. *American Educational Research Journal, 30*(2), 393–425.

Hill, H. C., Ball, D. L., & Schilling, S. G. (2008). Unpacking pedagogical content knowledge: Conceptualising and measuring teachers' topic-specific knowledge of students. *Journal for Research in Mathematics Education, 39*(4), 372–400.

Hodge, L., Visnovska, J., Zhao, Q., & Cobb, P. (2007). What does it mean for an instructional task to be effective? In J. Watson & K. Beswick (Eds.), *Mathematics: Essential research, essential practice* (Proceedings of the 30th annual conference of the Mathematics Education Research Group of Australasia, Vol. 1, pp. 392–401). Adelaide: MERGA.

Hollingsworth, H., Lokan, J., & McCrae, B. (2003). *Teaching mathematics in Australia: Results from the TIMSS video study (TIMSS Australia Monograph No. 5).* Camberwell, VIC: Australian Council for Educational Research.

Human Capital Working Group. (2008). *National numeracy review report.* Canberra: Author.

Keitel, C. (1993). Implicit mathematical models in social practice and explicit mathematics teaching by applications. In J. de Lange, C. Keitel, I. Huntley, & M. Niss (Eds.), *Innovation in mathematics education by modelling and applications* (pp. 138–151). Chichester: Ellis Horwood.

Kilpatrick, J., Swafford, J., & Findell, B. (Eds.). (2001). *Adding it up: Helping children learn mathematics.* Washington, DC: National Academy Press.

Kuhs, T. M., & Ball, D. L. (1986). *Approaches to teaching mathematics: Mapping the domains knowledge, skills, and dispositions.* East Lansing, MI: Michigan State University Center on Teacher Education.

Lampert, M. (2001). *Teaching problems and the problems of teaching.* New Haven, CT: Yale University Press.

Leder, G., Pehkonen, E., & Tőrner, G. (2002). *Beliefs: A hidden variable in mathematics education.* Dordrecht: Kluwer.

Leung, S. S. (1998). On the open-ended nature of mathematical problem solving. In E. Pehkonen (Ed.), *Use of open-ended problems in mathematics classrooms* (pp. 26–35). Helsinki: Department of Teacher Education, University of Helsinki.

Lovitt, C., & Clarke, D. M. (1988). *Mathematics Curriculum and Teaching Program activity bank* (Vol. 1). Carlton, VIC: Curriculum Corporation.

Lubienski, S. (2000). Problem solving as a means towards mathematics for all: An exploratory look through a class lens. *Journal for Research in Mathematics Education, 31*(4), 454–482.

Luke, A., Elkin, J., Weir, K., Land, R., Carrington, V., Dole, S., et al. (2003). *Beyond the middle: Executive summary.* St Lucia, QLD: University of Queensland.

Maier, E. (1991). Folk mathematics. In M. Harris (Ed.), *School, mathematics and work* (pp. 62–66). London: Palmer.

McLeod, D. B., & Adams, V. M. (1989). *Affect and problem solving.* New York: Springer.

Meyer, M., Dekker, T., Querelle, N., & Reys, R. (2001). Contexts in mathematics curriculum. *Mathematics Teaching in the Middle School, 6*(9), 522–527.

Middleton, J. A. (1995). A study of intrinsic motivation in the mathematics classroom: A personal construct approach. *Journal for Research in Mathematics Education, 26*(3), 254–279.

Middleton, J. A., & van den Heuvel-Panhuizen, M. (1995). The ratio table. *Mathematics Teaching in the Middle School, 1*(4), 282–288.

Middleton, J. A., & Jansen, A. (2011). *Motivation matters and interest counts.* Reston, VA: National Council of Teachers of Mathematics.

National Council of Teachers of Mathematics. (2000). *Principles and standards for school mathematics.* Reston, VA: NCTM

Nisbet, S., & Warren, E. (2000). Primary school teachers' beliefs relating to mathematics, teaching and assessing mathematics and factors that influenced those beliefs. *Mathematics Teacher Education and Development, 2*, 34–47.

Niss, M. (1987). Applications and modelling in the mathematics classroom: State and trends. *International Journal for Mathematical Education in Science and Technology, 18*(4), 487–505.

Pajares, M. F. (1992). Teachers' beliefs and educational research: Cleaning up a messy construct'. *Review of Educational Research, 6*(3), 307–332.

Peter-Koop, A. (2004). Fermi problems in primary mathematics classrooms: Pupils' interactive modelling processes. In I. Putt, R. Farragher, & M. McLean (Eds.), *Mathematics education for the third millennium: Towards 2010* (Proceedings of the 27th annual conference of the Mathematics Education Research Group of Australasia, pp. 454–461). Townsville, QLD: MERGA.

Philipp, R. A. (2007). Mathematics teachers' beliefs and affect. In F. K. Lester (Ed.), *Second handbook of research on mathematics teaching and learning* (pp. 257–315). Charlotte, NC: Information Age Publishing & National Council of Teachers of Mathematics.

Rowland, T., Huckstep, P., & Thwaites, A. (2005). Elementary teachers' mathematics subject knowledge: The knowledge quartet and the case of Naomi. *Journal of Mathematics Teacher Education, 8*(3), 255–281.

Ruthven, K., Laborde, C., Leach, J., & Tiberghien, A. (2009). Design tools in didactical research: Instrumenting the epistemological and cognitive aspects of the design of teaching sequences. *Educational Researcher, 38*(5), 329–342.

Shimizu, Y., Kaur, B., Huang, R., & Clarke, D. (Eds.). (2010). *Mathematical tasks in classrooms around the world*. Rotterdam: Sense Publishers.

Shulman, L. S. (1986). Those who understand: Knowledge growth in teaching. *Educational Researcher, 15*(2), 4–14.

Silver, E. A. (1994). On mathematical problem posing. *For the Learning of Mathematics, 14*(1), 19–28.

Simon, M. A. (1995). Reconstructing mathematics pedagogy from a constructivist perspective. *Journal of Research in Mathematics Education, 26*(2), 114–145.

Stacey, K., Helme, S., Archer, S., & Condon, C. (2001). The effect of epistemic fidelity and accessibility on teaching with physical materials: A comparison of two models for teaching decimal numeration. *Educational Studies in Mathematics, 47*(2), 199–221.

Staples, M. E. (2008). Promoting student collaboration in a detracked, heterogeneous secondary mathematics classroom. *Journal of Mathematics Teacher Education, 11*, 349–371.

Stein, M. K., Engle, R. A., Smith, M. S., & Hughes, E. K. (2008). Orchestrating productive mathematical discussions: Five practices for helping teachers move beyond show and tell. *Mathematical Thinking and Learning, 10*, 313–340.

Stein, M. K., Grover, B. W., & Henningsen, M. (1996). Building student capacity for mathematical thinking and reasoning: An analysis of mathematical tasks used in reform classrooms. *American Educational Research Journal, 33*(2), 455–488.

Stein, M. K., & Lane, S. (1996). Instructional tasks and the development of student capacity to think and reason and analysis of the relationship between teaching and learning in a reform mathematics project. *Educational Research and Evaluation, 2*(1), 50–80.

Sullivan, P. (1999). Seeking a rationale for particular classroom tasks and activities. In J. M. Truran & K. N. Truran (Eds.), *Making the difference* (Proceedings of the 21st annual conference of the Mathematics Educational Research Group of Australasia, pp. 15–29). Adelaide: MERGA.

Sullivan, P. (2011). *Teaching mathematics: Using research informed strategies*. Australian Education Review. Melbourne: Australian Council for Educational Research.

Sullivan, P., Clarke, D. M., Clarke, B. A., & O'Shea, H. (2009). Exploring the relationship between tasks, teacher actions, and student learning. In M. Tzekaki, M. Kaldrimidou, & H. Sakonidis (Eds.), *In search of theories in mathematics education* (Proceedings of the 33rd conference of the International Group of Psychology of Mathematics Education, Vol. 5, pp. 185–192). Thessaloniki, Greece: PME.

Sullivan, P., & Lilburn, P. (1997). *Open-ended maths activities: Using "good" questions to enhance learning*. Melbourne: Oxford University Press.

Sullivan, P., Mousley, J., & Zevenbergen, R. (2004). Describing elements of mathematics lessons that accommodate diversity in student background. In M. Johnsen Joines & A. Fuglestad (Eds.), *Proceedings of the 28th annual conference of the International Group for the Psychology of Mathematics Education* (pp. 257–265). Bergen: PME.

Sullivan, P., Mousley, J., & Zevenbergen, R. (2006). Developing guidelines for teachers helping students experiencing difficulty in learning mathematics. In P. Grootenboer, R. Zevenbergen & M. Chinnappan (Eds.), *Identities, cultures and learning spaces* (Proceedings of the 29th annual conference of the Mathematics Education Research Group of Australasia, pp. 496–503). Sydney: MERGA.

Sullivan, P., Prain, V., Campbell, C., Deed, C., Drane, S., Faulkner, M., et al. (2009). Trying in the middle years: Students' perceptions of their aspirations and influences on their efforts. *Australian Journal of Education, 5*(2), 176–191.

Sullivan, P., Zevenbergen, R., & Mousley, J. (2002). Contexts in mathematics teaching: Snakes or ladders. In B. Barton, K. Irwin, M. Pfannkuch, & M. Thomas (Eds.), *Mathematics education in the South Pacific* (Proceedings of the 25th annual conference of the Mathematics Education Group of Australasia, pp. 649–656). Sydney: MERGA.

Swan, M. (2005). *Improving learning in mathematics: Challenges and strategies*. London: Department for Education and Skills Standards Unit.

Swan, M. (2006). *Collaborative learning in mathematics: A challenge to our beliefs and practices*. Leicester: National Institute of Adult Continuing Education.

Thompson, A. G. (1992). Teachers' beliefs and conceptions: A synthesis of research. In D. A. Grouws (Ed.), *Handbook of research on mathematics teaching and learning* (pp. 127–146). New York: Macmillan.

Thorndike, E. L. (1926). *The psychology of arithmetic*. New York: Macmillan.

Tschannen-Moran, M., Hoy, A., & Hoy, W. (1998). Teaching efficacy: Its meaning and measure. *Review of Educational Research, 68*(2), 202–248.

Tzur, R. (2008). A researcher perplexity: Why do mathematical tasks undergo metamorphosis in teacher hands? In O. Figuras, J. L. Cortina, S. Alatorre, T. Rojano, & A Sepulveda (Eds.), *Proceedings of the 32nd annual conference of the International Group for the Psychology of Mathematics Education* (pp. 1–139, 147). Morelia: PME.

van den Heuvel-Panhuizen, M. (2005). The role of contexts in assessment problems in mathematics. *For the Learning of Mathematics, 25*(2), 2–9.

Vygotsky, V. (1978). *Mind in society*. Cambridge, MA: Harvard University Press.

Watson, A., & Mason, J. (1998). *Questions and prompts for mathematical thinking*. Derby: Association of Teachers of Mathematics.

Watson, A., & Sullivan, P. (2008). Teachers learning about tasks and lessons. In D. Tirosh & T. Wood (Eds.), *Tools and resources in mathematics teacher education* (pp. 109–135). Rotterdam: Sense Publishers.

Wiliam, D. (1998, July). *Open beginnings and open ends*. Paper distributed at the Open-Ended Questions Discussion Group, International Conference for the Psychology of Mathematics Education, Stellenbosch, South Africa.

Zan, R., & di Martino, P. (2010). "Me and maths": Toward a definition of attitude grounded on students' narrative. *Journal of Mathematics Teacher Education, 13*(1), 27–48.

Zevenbergen, R., & Zevenbergen, K. (2009). The numeracies of boatbuilding: New numeracies shaped by workplace technologies. *International Journal of Science and Mathematics Education, 7*(1), 183–206.

Index

P. Sullivan et al., *Teaching with Tasks for Effective Mathematics Learning*, Mathematics Teacher Education 9, DOI 10.1007/978-1-4614-4681-1,

U

Z

MIX
Papier aus verantwortungsvollen Quellen
Paper from responsible sources
FSC® C105338

If you have any concerns about our products,
you can contact us on
ProductSafety@springernature.com

In case Publisher is established outside the EU,
the EU authorized representative is:
Springer Nature Customer Service Center GmbH
Europaplatz 3, 69115 Heidelberg, Germany

Printed by Libri Plureos GmbH
in Hamburg, Germany